# TIME, CHANCE AND REDUCTION

## Philosophical Aspects of Statistical Mechanics

Statistical mechanics attempts to explain the behaviour of macroscopic physical systems in terms of the mechanical properties of their constituents. Although it is one of the fundamental theories of physics, it has received little attention from philosophers of science. Nevertheless, it raises philosophical questions of fundamental importance on the nature of time, chance and reduction. Most philosophical issues in this domain relate to the question of the reduction of thermodynamics to statistical mechanics.

This book addresses issues inherent in this reduction: the time-asymmetry of thermodynamics and its absence in statistical mechanics; the role and essential nature of chance and probability in this reduction when thermodynamics is non-probabilistic; and how, if at all, the reduction is possible. Containing contributions on current research by experts in the field, this is an invaluable survey of the philosophy of statistical mechanics for academic researchers and graduate students interested in the foundations of physics.

GERHARD ERNST is a Professor of History of Philosophy and Moral Philosophy at Universität Stuttgart. His main research interests are in moral philosophy, epistemology and philosophy of science.

ANDREAS HÜTTEMANN is a Professor of Philosophy at Westfälische Wilhelms-Universität Münster. His research interests include philosophy of science and early modern philosophy.

# TIME, CHANCE AND REDUCTION

## Philosophical Aspects of Statistical Mechanics

*Edited by*

GERHARD ERNST
*Universität Stuttgart*

ANDREAS HÜTTEMANN
*Westfälische Wilhelms-Universität Münster*

CAMBRIDGE UNIVERSITY PRESS
Cambridge, New York, Melbourne, Madrid, Cape Town, Singapore, São Paulo, Delhi

Cambridge University Press
The Edinburgh Building, Cambridge CB2 8RU, UK

Published in the United States of America by Cambridge University Press, New York

www.cambridge.org
Information on this title: www.cambridge.org/9780521884013

© Cambridge University Press 2010

This publication is in copyright. Subject to statutory exception
and to the provisions of relevant collective licensing agreements,
no reproduction of any part may take place without the written
permission of Cambridge University Press.

First published 2010

Printed in the United Kingdom at the University Press, Cambridge

*A catalogue record for this publication is available from the British Library*

ISBN 978-0-521-88401-3 Hardback

Cambridge University Press has no responsibility for the persistence or
accuracy of URLs for external or third-party internet websites referred to
in this publication, and does not guarantee that any content on such
websites is, or will remain, accurate or appropriate.

# Contents

|   |   |   |
|---|---|---|
| | *List of contributors* | *page* vi |
| 1 | Introduction | 1 |
| | *Gerhard Ernst and Andreas Hüttemann* | |

**Part I: The arrows of time**    11

2  Does a low-entropy constraint prevent us from influencing the past?    13
*Mathias Frisch*

3  The past hypothesis meets gravity    34
*Craig Callender*

4  Quantum gravity and the arrow of time    59
*Claus Kiefer*

**Part II: Probability and chance**    69

5  The natural-range conception of probability    71
*Jacob Rosenthal*

6  Probability in Boltzmannian statistical mechanics    92
*Roman Frigg*

7  Humean metaphysics versus a metaphysics of powers    119
*Michael Esfeld*

**Part III: Reduction**    137

8  The crystallization of Clausius's phenomenological thermodynamics    139
*C. Ulises Moulines*

9  Reduction and renormalization    159
*Robert W. Batterman*

10  Irreversibility in stochastic dynamics    180
*Jos Uffink*

*Index*    208

# Contributors

ROBERT W. BATTERMAN Department of Philosophy, Talbot College, University of Western Ontario, London ON, N6A 3K7, Canada

CRAIG CALLENDER Philosophy Department, University of California San Diego, 9500 Gilman Drive, La Jolla CA 92093–0119, USA

GERHARD ERNST Institut für Philosophie, Universität Stuttgart, Seidenstraße 36, D-70174 Stuttgart, Germany

MICHAEL ESFELD Université de Lausanne, Section de Philosophie, CH-1015 Lausanne, Switzerland

ROMAN FRIGG Department of Philosophy, Logic and Scientific Method, London School of Economics and Political Science, Houghton Street, London WC2A 2AE, UK

MATHIAS FRISCH Skinner Building, University of Maryland, College Park MD 20742, USA

ANDREAS HÜTTEMANN Philosophisches Seminar, Westfälische Wilhems-Universität Münster, Domplatz 23, D-48143 Münster, Germany

CLAUS KIEFER Institut für Theoretische Physik, Universität zu Köln, Zülpicher Straße 77, D-50937 Köln, Germany

C. ULISES MOULINES Seminar für Philosophie, Logik und Wissenschaftstheorie, Ludwig-Maximilians-Universität München, Ludwigstraße 31, D-80539 München, Germany

JACOB ROSENTHAL Institut für Philosophie der Universität Bonn, Am Hof 1, D-53113 Bonn, Germany

JOS UFFINK Institute for History and Foundations of Science, Utrecht University, P.O. Box 80.000, NL-3508 TA, Utrecht, The Netherlands

# 1
# Introduction

GERHARD ERNST AND ANDREAS HÜTTEMANN

## 1.1 Statistical mechanics and philosophy

Statistical mechanics attempts to explain the behaviour of macroscopic physical systems (in particular their thermal behaviour) in terms of the mechanical properties of their constituents. In order to achieve this aim it relies essentially on probabilistic assumptions. Even though in general we do not know much about the detailed behaviour of each degree of freedom (each particle), statistical physics allows us to make very precise predictions about the behaviour of systems such as gases, crystals, metals, plasmas, magnets as wholes.

The introduction of probabilistic concepts into physics by Maxwell, Boltzmann and others was a significant step in various respects. First, it led to a completely new branch of theoretical physics. Second, as Jan von Plato pointed out, the very meaning of probabilistic concepts changed under the new applications. To give an example: whereas before the development of statistical physics variation could be conceived as the deviation from an ideal value this was no longer a tenable interpretation in the context of statistical physics. Genuine variation had to be accepted (von Plato, 2003: 621).

Furthermore, the introduction of probabilistic concepts triggered philosophical speculations, in particular with respect to the question whether the atomic world does indeed follow strict deterministic laws (cf. von Plato, 1994; Stöltzner, 1999). For instance, in 1873 Maxwell gave a lecture entitled 'Does the Progress of Physical Science tend to give any advantage to the opinion of Necessity (or Determinism) over that of the Contingency of Events and the Freedom of the Will?' He wondered whether 'the promotion of natural knowledge may tend to remove the prejudice in favour of determinism which seems to arise from assuming that the physical science of the future is a mere magnified image of that of the past' (quoted in von Plato, 1994: 87). Other physicists in the field voiced similar views. Franz Exner, an Austrian physicist, argued in his lectures on the physical foundations of natural science that the concepts of causation and laws of nature have to be revised in the light of the need to introduce probabilistic concepts into physics. With respect to determinism he claims that deterministic laws obtain only in the macro-world and that there is no reason to assume that determinism is true at the level of atoms. It may very well be the case, he argues, that the most fundamental laws of physics are probabilistic. (The lectures were written in and before the first world war; Exner, 1922: lectures 86–95.) These views which arose in the context of philosophical reflections on the nature of the new statistical mechanics may well have contributed to the willingness with which some proponents of the newly emerging quantum mechanics gave up determinism.

Be that as it may, philosophical reflection on statistical mechanics came almost to an end with the advent of other new fundamental physical theories, such as quantum mechanics and the theory of general relativity. For sixty years philosophers of physics focused almost exclusively on the interpretation of quantum mechanics and the philosophical implications of the theory of general relativity. It was only with the publication of Lawrence Sklar's *Physics and Chance* (1993) that the discussion of philosophical and foundational problems of statistical mechanics became more popular again among philosophers of physics. Sklar (1993: 6) surmised that the neglect of statistical mechanics was partly due to the fact that the field itself is in a certain disarray. Compared to the situation in quantum mechanics and the theories of relativity there are not only different philosophical interpretations of physical theories but also widely diverging approaches and schools within statistical mechanics itself. It was Sklar's aim to provide a survey and make some of these approaches accessible and thus to stimulate further philosophical investigations. It seems that by and large he succeeded. The last decade has seen the field flourishing. There have been major philosophical monographs like David Albert's *Time and Chance* (2000) as well as a lot of other work – to a significant extent by the contributors of this volume.[1] There is even a recent volume entitled *Contemporary Debates in Philosophy of Science* (by Christopher Hitchcock, 2004), in which philosophical issues pertaining to statistical mechanics figure more prominently than those of quantum mechanics or the theories of relativity.

The main philosophical and foundational questions that are currently discussed concern the relation of statistical mechanics and thermodynamics. Thermodynamics started as a theory of heat engines. It gradually developed into a rather general theory that describes matter in all its phases and their thermal and magnetic behaviour in particular. Thermodynamics makes virtually no assumptions about the micro-structure of the systems it describes. But of course the systems described by thermodynamics do have a micro-structure. Statistical mechanics was developed in

---

[1] For a very helpful state of the art article see Uffink (2007).

the hope to explain the thermodynamic macro-laws in terms of the behaviour of the systems' constituents. Surprisingly, this reductive enterprise turned out to be rather difficult. Most foundational/philosophical issues are related in one way or another to the question of the reduction of thermodynamics to statistical mechanics. In particular, there are three focal issues:

(1) One law in thermodynamics exhibits a pertinent time-asymmetry (according to the second law of thermodynamics, the entropy in a closed system never decreases in time), whereas the fundamental laws of statistical mechanics fail to exhibit such an asymmetry. So, where does the time-asymmetry come in? And how is the thermodynamic time-asymmetry related to other 'arrows of time' in physics?
(2) Statistical mechanics is a probabilistic theory, while thermodynamics is not. So, how is the concept of probability to be interpreted in order to be coherent and to make thermodynamic behaviour intelligible?
(3) Thermodynamics allegedly is reducible to statistical mechanics. But how exactly, if at all, does the reduction work? And what concept of reduction is here employed anyway?

Part one of this collection of essays is concerned mainly with the first topic (see Section 1.2), part two with the second (see Section 1.3), and part three with the third (see Section 1.4).

## 1.2 The arrows of time

Time seems to have a direction. However, it is not so clear what this claim exactly amounts to and whether a direction *of* time could be distinguished from processes *in* time having a direction. Even though the fundamental dynamical laws in physics do not have a temporal direction both in physics as well as in ordinary life we come across various temporally directed phenomena ('arrows of time'). There are several such arrows in physics (see the chapter by Kiefer): We only observe certain sorts of radiation, entropy never decreases and our universe expands. Similarly, causation seems to have a temporal direction. Furthermore, we seem to know more about the past than about the future. And counterfactual dependence seems to be temporally directed as well: 'If A had not occurred then C would not have occurred either' seems to be – in general – a good candidate for a true proposition only if what is described by A precedes what is described by C (see Horwich, 1987). The chapters by Mathias Frisch, Craig Callender and Claus Kiefer are all concerned with the arrows of time.

The contribution by Mathias Frisch deals with the question of how various of the temporal asymmetries are related to one another. There is a long tradition according to which the causal asymmetry is closely related to the temporal asymmetry embodied in the second law of thermodynamics. This view has recently been defended by David Albert and by Barry Loewer, who argue that the causal asymmetry can be grounded in those facts that explain the second law of thermodynamics.

Both accounts centrally involve the claim that it follows from Boltzmann's account of the thermodynamic asymmetry (the so-called past hypothesis, which postulates a low-entropy constraint on the early universe) that possible macro-evolutions are much more restricted toward the past than toward the future. The statistical mechanical account, as Loewer puts it, results in a time-asymmetric 'tree-structure' for possible macro-evolutions. Frisch argues that statistical mechanics allows not only for branchings but also for the reconvergence and merging of possible macro-histories. As a consequence he maintains that Albert's and Loewer's accounts do not work in their present form because they fail to explain how our *strict* concepts of causal influence and causal control emerge.

Craig Callender's chapter argues that it is essential for answering the problem of the direction of time to take gravity into account. More particularly, he deals with the question whether the low-entropy constraint is plausible given what we know about gravity. The past hypothesis, which is invoked to explain why entropy increases (or rather: never decreases) seems to be *prima facie* patently false. According to current cosmological theories the early universe is an almost homogeneous isotropic state of approximately uniform temperature, i.e. a very high entropy state, not a low-entropy state as postulated by the past hypothesis. The standard response to this objection is that we forgot to include gravity. Gravity, it is said, saves the past hypothesis. So now the essential question is whether the gravitational behaviour of the stars etc. can plausibly be interpreted as the movement to an equilibrium state. Callender argues that the inclusion of gravity into the Boltzmannian account of the direction of time is highly non-trivial. After sketching some serious problems with gravity, he develops a sketch of how one can obtain a never decreasing Boltzmann entropy in self-gravitating systems described by certain types of gravitational kinetic equations.

Claus Kiefer approaches the different arrows of time from the perspective of a physicist. His aim is to explain how the various physical arrows can in principle be understood on the basis of a fundamental theory of quantum gravity. After a brief discussion of the time-directed processes in question, the physical framework of a particular approach to quantum gravity, the canonical approach, is outlined. The fundamental equation (the Wheeler–DeWitt equation) is devoid of any classical time parameter, but involves a new type of dynamics with respect to an intrinsic time. In simple models this intrinsic time is given by the 'radius' of the universe. Standard time can be recovered in certain situations as an approximation. Kiefer claims that given a natural boundary condition on the Wheeler–DeWitt equation, an arrow of time follows automatically.

## 1.3 Probability and chance

Probabilistic reasoning is essential for statistical mechanics and discussions of the foundations of statistical mechanics often focus on how to justify particular probabilistic assumptions. There is, however, a problem that is systematically prior: What are these probabilities? Subjectivists argue that probabilities reflect our ignorance of the true state of the system. Objectivists deny this and submit that probabilities have to be objective, i.e. that they have to be chances.

It is now often assumed that the probabilities introduced into statistical mechanics should not be given an epistemic or subjectivist reading. David Albert, for instance, comments on the subjectivist approach: 'Can anybody seriously think that [our epistemic situation, A. H.] would somehow *explain* the fact that the *actual microscopic conditions of actual thermodynamic systems are statistically distributed in the way that they are*? Can anybody seriously think that it is somehow *necessary*, that it is somehow *a priori*, that the particles that make up the material world must arrange themselves in accord with *what we know*, with what *we happen to have looked into*? Can anybody seriously think that our merely being *ignorant* of the exact microconditions of thermodynamic systems plays some part in *bringing it about*, in *making it the case, that (say) milk dissolves in coffee*? How could that be?' (Albert, 2000: 64). The chapters in the second part of this volume focus on the interpretation of probabilities and related metaphysical issues.

Jacob Rosenthal and Roman Frigg discuss objectivist approaches to probability. Usually two versions of this approach are considered, the frequency interpretation of probability and the propensity interpretation. Both of these face serious challenges – as interpretations of probabilities in general (see Rosenthal's chapter for a brief review) as well as for physical reasons when applied to probabilistic reasoning in statistical mechanics (see Section 6.2 of Frigg's chapter). Jacob Rosenthal defends a third objectivist interpretation of probability, the natural-range conception of probability. It considers probabilities as deriving from ranges in suitably structured initial state spaces. Roughly, the probability of an event is the proportion of initial states that lead to this event in the space of all possible initial states, provided that this proportion is approximately the same in any not too small interval of the initial-state space. The range approach to probabilities is usually treated as an *explanation* for the occurrence of probabilistic patterns, whereas Rosenthal examines its prospects for an objective *interpretation* of probability, in the sense of providing truth conditions for probability statements that do not depend on our state of mind or information. The main objection to such a proposal is that it is circular, i.e. presupposes the concept of probability, because a measure on the initial state has to be introduced. Rosenthal argues that this objection can be successfully met and that the range approach has better prospects to provide a satisfactory objective interpretation of probability statements than frequency or propensity accounts.

Roman Frigg focuses on the question of how different approaches to statistical mechanics introduce probabilities and what kind of interpretation of probabilities has to be assumed. He confines himself to approaches that follow Boltzmann in defining an entropy function $S_{B(t)}$ in terms of the micro-state of the system. Different approaches in this tradition diverge in how they introduce probabilities into the theory and in how they explain the tendency of $S_{B(t)}$ to increase. The most fundamental distinction is between approaches that assign probabilities directly to the system's macro-states ('macro-probabilities'), and approaches that assign probabilities to the system's micro-state being in particular subsets of the macro-region corresponding to the system's current macro-state ('micro-probabilities'). More particularly, Frigg discusses Boltzmann's own proposal to assign probabilities to the system's macro-states and the view by Paul and Tatiana Ehrenfest that these should be interpreted as time averages, presupposing the ergodicity of the system. It is now well known that this approach faces serious difficulties. Frigg's point is that even if these were surmounted there would be a grave problem not so much with the probabilistic assumptions as such but rather with the dynamical laws that are supposed to bring about the increase in entropy. The micro-probabilistic alternative discussed, has been proposed by David Albert (2000). Frigg discusses this approach and Barry Loewer's suggestion for understanding these probabilities as Humean chances in David Lewis's sense. Frigg reaches the same conclusion as in the discussion of macro-probabilities. All that is needed to *explain* why things happen is the initial condition and the dynamics. Frigg opts for an epistemic interpretation of probabilities in statistical mechanics. Probabilities have no role to play in *explaining* why a system behaves as it does. They are introduced for reasons of epistemic limitations.

The chapter by Michael Esfeld deals with metaphysical issues. In particular, he is concerned with the opposition between Humean metaphysics and the metaphysics of powers. Esfeld argues that within the bounds of Humean metaphysics everything is a matter of contingency. Consequently, there is no deep metaphysical difference between a deterministic world and a world in which only probabilistic laws hold. This position is contrasted with the foundations of probabilities according to the metaphysics of powers, in particular with the view that traces probabilities back to propensities. Esfeld discusses arguments for and against both positions. In the end he opts for the metaphysics of powers, mainly for two reasons: (1) the metaphysics of powers avoids certain troublesome commitments; (2) contrary to a widespread belief, the metaphysics of powers is compatible with physics, and it is able to provide a complete and coherent ontology that does justice to both physics and the special sciences.

## 1.4 Reduction

There are many meanings of the term 'reduction'. The shared background for most discussions in philosophy of science is Nagel's account of the formal criteria of reduction (Nagel, 1961: ch. 11). Nagel's primary concern was whether an older theory (the reduced theory) is reducible to its successor (the reducing theory). Reduction was conceived of as a special case of deductive-nomological explanation. Successful 'Nagel reduction' integrates the old theory into the successor theory and provides a clear sense in which the successor theory is better than its predecessor. Ironically, Nagel presented the relation of thermodynamics to statistical mechanics as a paradigm case for a successful reduction in the sense he developed. In fact, this reduction is a highly non-trivial affair (cf. Ernst, 2003).

It is helpful to distinguish various senses of reduction. First, there is *diachronic* reduction. Diachronic reduction concerns the evolution of disciplines and their theories. This is the aspect Nagel had in mind when he introduced his account. This perspective can be contrasted with *synchronic* reduction. Synchronic reduction concerns the relation of two theories that pertain to the same realm. At least two cases of synchronic reduction in this sense can be distinguished. First, the two theories in question may be related as limiting cases. Robert Batterman has developed this view in his *The Devil in the Details* (2002) and works with it in his contribution (see below). Second, it may be the case that one theory describes the behaviour of compound systems (in a certain terminology) and the other theory describes the behaviour of the constituents of the systems (in a different terminology). In these cases we want to know how the different theoretical accounts fit together. It seems plausible that in this case reduction is required for reasons of coherence. Interestingly, if it turns out to be difficult to bring about coherence it is not necessarily the higher-level assumptions that are put into question. In the case of thermodynamics and statistical mechanics it is assumptions about micro-states that turned out to be most controversial.

The contribution by Ulises Moulines concerns diachronic reduction. Moulines distinguishes four different types of diachronic structures in the evolution of scientific disciplines: (a) the *evolution* of a theory; (b) the *replacement* of one theory by another; (c) the *embedding* of one theory into another; (d) the (slow) *crystallization* of a theory out of different previous elements. He argues that a typical example of crystallization is provided by the gradual emergence of phenomenological thermodynamics in the middle of the nineteenth century. A major role in this process was played by several writings of Rudolf Clausius. Two of them are analysed in the present chapter: *Über die bewegende Kraft der Wärme* (1850) and *Über eine veränderte Form des zweiten Hauptsatzes der mechanischen Wärmetheorie* (1854). By employing some formal tools of the structuralist reconstruction methodology (in particular, the notions of *model* and *theory-net*), three different theories are identified in Clausius' papers, thereby showing how much they still owe to the caloric theory, how much they contain germs of future developments of thermodynamics (such as the notion of *internal energy* or the difference between *reversible* and *irreversible cycles*), and finally how there are conceptual tensions in the interrelationships between Clausius' own theories.

Robert Batterman considers the so-called reduction of thermodynamics to statistical mechanics from both historical and relatively contemporary points of view. Today, most philosophers of physics doubt that the relation of the two theories can be described as a reduction in the sense of Nagel. Batterman turns to J. Willard Gibbs who can be seen as sharing the scepticism about the possibility of such a philosophical reduction of thermodynamics to statistical mechanics. Gibbs' account is not only discussed for his caution in connecting thermodynamical concepts with those from statistical mechanics. Batterman takes him to suggest that one should look for how thermodynamic quantities emerge from statistical quantities when certain limiting conditions are satisfied. The paper then develops the idea that the limit of statistical mechanics, as the number of degrees of freedom goes to infinity, should yield the continuum thermodynamic theory. Batterman sketches a program for reductive relations that involves deep connections between results in probability theory on limit theorems and the so-called real space renormalization techniques that play such an essential role in understanding the universality of critical phenomena. It turns out that this kind of (feasible) reduction yields only an association of thermodynamic properties, such as temperature and entropy, and a so-called universality class of statistical mechanical structures. This contrasts with Nagel-reduction, which requires (according to the standard reading) a one-to-one association of thermodynamic and statistical mechanical quantities.

Jos Uffink discusses the problem of explaining the emergence of irreversible processes in classical statistical mechanics from the point of view of stochastic dynamics – and thus the question of synchronic reduction in the second sense, i.e. concerning the relation of the behaviour of compound systems to that of their constituents. An influential approach to the foundations of classical non-equilibrium statistical mechanics claims to obtain a satisfactory explanation of irreversible behaviour by characterizing the evolution of macroscopic physical systems as a Markov process, or more abstractly, in terms of a semigroup of non-invertible evolution operators. The general formalism developed in this approach, sometimes called 'stochastic dynamics', can in fact be obtained from a variety of physical motivations, e.g. by assuming that the physical system interacts with an environment (the 'open systems' or 'interventionist' approach) or that only a few macroscopic variables from the detailed microscopic state of the state are relevant for its physical description ('coarse graining'). Uffink argues that, despite appearances, the usual assumptions of this approach remain fully time-reversal invariant, and hence do not embody irreversible behaviour. It is argued that a proper account of irreversibility should not focus on the Markov property but on a different definition of reversibility for stochastic processes.

## Acknowledgements

We would like to thank the *Junge Akademie an der Berlin-Brandenburgischen Akademie der Wissenschaften und der Deutschen Akademie der Naturforscher Leopoldina* (Berlin) for financial support of a workshop that took place in Munich in 2006. Most of the papers in this volume were originally presented on that occasion. In particular we would like to thank the members of the *AG Zufall, Zeit und Zustandssumme* for their help in organizing the workshop and Matthias Hoesch who provided invaluable help in preparing the manuscripts for publication.

Gerhard Ernst
Andreas Hüttemann

## References

Albert, D. (2000). *Time and Chance*. Cambridge, MA: Harvard University Press.
Batterman, R. (2002). *The Devil in the Details*. Oxford: Oxford University Press.
Ernst, G. (2003). *Die Zunahme der Entropie. Eine Fallstudie zum Problem nomologischer Reduktion*. Paderborn: Mentis.
Exner, F. (1922). *Vorlesungen über die physikalischen Grundlagen der Naturwissenschaften*. Vienna: Franz Deuticke.
Hitchcock, C. (2004). *Current Debates in Philosophy of Science*. Oxford: Oxford University Press.
Horwich, P. (1987). *Asymmetries in Time*. Cambridge, MA: MIT Press.
Nagel, E. (1961). *The Structure of Science*. London: Routledge.
Sklar, L. (1993). *Physics and Chance*. Cambridge: Cambridge University Press.
Stöltzner, M. (1999). Vienna Indeterminism: Mach, Boltzmann, Exner. *Synthese*, **119**, 85–111.
Uffink, J. (2007). Compendium of the Foundations of Classical Statistical Physics. In *Philosophy of Physics*, ed. J. Butterfield and J. Earman. Amsterdam: North-Holland, pp. 923–1047.
Von Plato, J. (1994). *Creating Modern Probability*. Cambridge: Cambridge University Press.
Von Plato, J. (2003). The rise of probabilistic thinking. In: *The Cambridge History of Philosophy 1870–1945*, ed. T. Baldwin. Cambridge: Cambridge University Press, pp. 621–628.

# Part I

The arrows of time

# 2

# Does a low-entropy constraint prevent us from influencing the past?

MATHIAS FRISCH

## 2.1 Introduction

It is part of our common sense conception of the world that what happens now can make a difference to the future but not to the past; events in the present, we believe, can causally influence the occurrence of future events but not of past events. What is the relation between this asymmetry and other physical asymmetries? Is the causal asymmetry fundamental or can our asymmetric notion of cause be shown to be reducible to some other physical asymmetry? There is a venerable tradition in the foundations of physics and the philosophy of science according to which the causal asymmetry is intimately related to the temporal asymmetry embodied in the second law of thermodynamics. This view has recently been forcefully defended by David Albert (2000) and by Barry Loewer (2007), who argue that the causal asymmetry can ultimately be grounded in the very same facts that give rise to the second law of thermodynamics, chiefly among them a low-entropy constraint on the initial state of the universe.

In this chapter I will critically examine aspects of their accounts and will argue that neither account is successful as developed so far.[1] In Section 2.2 I will briefly summarize the Boltzmannian account of the thermodynamic asymmetry, from which Albert and Loewer aim to derive asymmetries of causal influence and control. Both accounts centrally involve the claim that it follows from the Boltzmannian account that possible macro-evolutions are much more restricted toward the past than toward the future. The statistical mechanical account, as Loewer puts it, results in a time-asymmetric 'tree-structure' for possible macro-evolutions. I will criticize this claim in Section 2.3.

---

[1] For discussions of how a third asymmetry – the asymmetry of radiation – may be related to the asymmetries of thermodynamics and of causation, see Frisch (2000; 2005; 2006b).

While there is a considerable amount of agreement between the two accounts, Albert and Loewer emphasize different routes by which the thermodynamic asymmetry is meant to ground our time-asymmetric notions of causal influence or control. Albert, in the first instance, focuses on a purely physical asymmetry that he takes to be exhibited by suitably small and localized macro-events of the kind that figure in paradigmatic causal judgments (such as that the collision between two billiard balls caused one ball to roll into the corner pocket). In Albert's terminology, such macro-events can provide us with *causal handles* on the future but not on the past. Loewer's primary focus, by contrast, is on human actions and a putative asymmetry characterizing *decision counterfactuals*, whose antecedents make reference to possible decisions. In Sections 2.4 and 2.5 I will raise worries about both of these routes of trying to connect up the statistical mechanical asymmetry to the asymmetry of control or influence.

Albert and Loewer's aim is to offer an account of how it is that we possess a time-asymmetric concept of causal influence or control by showing that such a concept tracks certain non-causal features of the world given by fundamental physics. Common sense causal claims are by and large concerned with relatively small, 'human-sized' macro events of the kind that could be the result of human interventions. Arguably, then, any account of how we come to possess an asymmetric concept of cause need only be able to reproduce the asymmetry as far as causal claims within this domain are concerned.[2] It is, however, part of our notion of causal relations among 'human-sized' macro-events that the temporal asymmetry of causation is strict in the sense that in *all* paradigmatic or standard circumstances the relation of causation is future directed and that in such cases there is absolutely *no* backward causation. We believe that our interventions can have an effect on the future development of the world and we also believe that our interventions can have *absolutely no* effect on the past. A successful entropy account would have to be able to account for this feature of our concept. This does not mean that it has to be a consequence of the account that there is no backward causal influence. Since entropy accounts ultimately appeal to certain probabilistic relations that they take to be derivable from statistical mechanics, they may have the consequence that the causal asymmetry is not strict. Nevertheless, the account has to be able to explain why *we take* the asymmetry to be strict.[3] Thus, similar to derivations

---

[2] Thus, Albert rightly argues that the fact that a universe in the shape of Bozo the clown would have to have had a very different past from ours would not undermine his entropy-account of causation (see Albert 2000: 130, fn. 21). Even if Albert's account had the consequence that there is backward causal dependence in this case, this will pose no problem for his account, since this is not the kind of case that could have played a role in our acquisition of causal concepts.

[3] I think it may even be compatible with our common sense notion of causation, that there could be arcane-physical circumstances in which there is backward causation. The point I am making here is that we take the asymmetry to be strict as far as common sense, 'billiard ball-like' circumstances are concerned.

of the second law from statistical consideration it would have to be shown that in paradigmatically causal contexts exceptions to the asymmetry are extremely rare and improbable. It would be a problem for an entropy account, if it implied that there will be widespread backward causal influence in the kind of circumstances that we take to be paradigmatically forward causal.

## 2.2 The micro-statistical account

According to the Boltzmannian account defended in Albert (2000), the thermodynamic asymmetry that the entropy of a closed macroscopic system never decreases can be explained by appealing to a time-symmetric micro-dynamics and an asymmetric constraint on initial conditions. If we assume an equiprobability distribution of micro-states compatible with a given macro-state of non-maximal entropy, then it can be made plausible that, intuitively, most micro-states will evolve into states corresponding to macro-states of higher entropy. However, if the micro-dynamics governing the system is time-symmetric, then the same kind of considerations also appear to show that, with overwhelming probability, the system evolved *from* a state of higher entropy. This undesirable retrodiction, which is at the core of the *reversibility objection*, can be blocked, if we conditionalize the distribution of micro-states not on the present macro-state but on a low-entropy initial state of the system. Since the reversibility objection can be raised for any time in the past as well, Albert and others argue that we are ultimately led to postulate an extremely low-entropy state at or near the beginning of the universe. Albert calls this postulate 'the past hypothesis' (*PH*).

Positing an equiprobability distribution at some initial time, however, seems to lead to the following problem. If we postulate a uniform probability distribution over the initial state of a system, then the distribution will not be uniform over micro-states compatible with the actual macro-state at later times. If later macro-states have higher entropy, they will correspond to regions of phase space that are vastly larger than the region corresponding to the low-entropy initial state. But, according to Liouville's theorem, regions of phase space evolve into regions of equal size. Thus, positing an equiprobability distribution at the initial time precludes that the distribution is uniform over macro-states at later times and, hence, might seem to preclude us from bringing to bear the very considerations that seemed to ensure that entropy is overwhelmingly likely to increase in the first place.

This problem can be solved, if we assume that the phase space region corresponding to the initial macro-state dynamically evolves into a highly fibrillated region such that the micro-states that have evolved from the initial macro-state eventually are homogeneously distributed over all measurable subregions of the system's available phase space. A formal condition that ensures that this assumption is the condition that a system is *mixing* (see, e.g., Uffink, 2006). A dynamical system is a tuple $\langle \Gamma, A, \mu, T \rangle$, where $\Gamma$ is the system's phase space, $A$ is the set of measurable subsets of $\Gamma$, $\mu$ is a probability measure, and $T$ is a one parameter group of transformations $T_t$ that represents the evolution operators. A dynamical system is mixing, just in case, for all $A, B \in A$

$$\lim_{t \to \infty} \mu(T_t A \cap B) = \mu(A)\mu(B)$$

For such a system, the micro-state at $t$ will with overwhelming probability be 'typical' of the micro-states compatible with the macro-state at $t$, in the sense required for the Boltzmannian account.

Thus, the assumptions of the statistical mechanical account (SM) from which the thermodynamic asymmetry is derived are the following:

(i) time-symmetric, deterministic dynamical micro-laws;
(ii) the *past hypothesis PH*, which characterizes the initial macro-state of the universe as a low-entropy condition satisfying certain further symmetry conditions;
(iii) a *probability postulate PROB*, which postulates a uniform probability distribution over the physically possible initial micro-states of the universe, compatible with the past hypothesis *PH*;
(iv) an assumption of mixing or dynamic instability of possible micro-evolutions.

This account of the entropy-asymmetry has been challenged (e.g. by Winsberg, 2004; Earman, 2006), but I do not want to discuss these criticisms here. In what follows I will assume that the account can successfully explain the thermodynamic asymmetry and ask whether or not it can be extended to explain an asymmetry of influence as well.

## 2.3 Trees or webs

Immediately after his own summary of how the micro-statistical account appealing to *PH* and *PROB* accounts for the entropy asymmetry, Loewer presents the following figure (Fig. 2.1), which, he says, 'is a depiction of possible evolutions of micro- and macro-states that should provide an idea of how all this goes' (Loewer, 2007: 300).

In this figure thin lines represent possible micro-histories of the universe and thick cylinders represent possible macro-histories. Possible macro-histories, Loewer claims, exhibit a tree structure: even though the evolution of micro-histories is assumed to be deterministic, the evolution of macro-histories is future-indeterministic in that more than one future macro-history will in general be compatible with the macro-state of the world at a time. Indeed, there are different

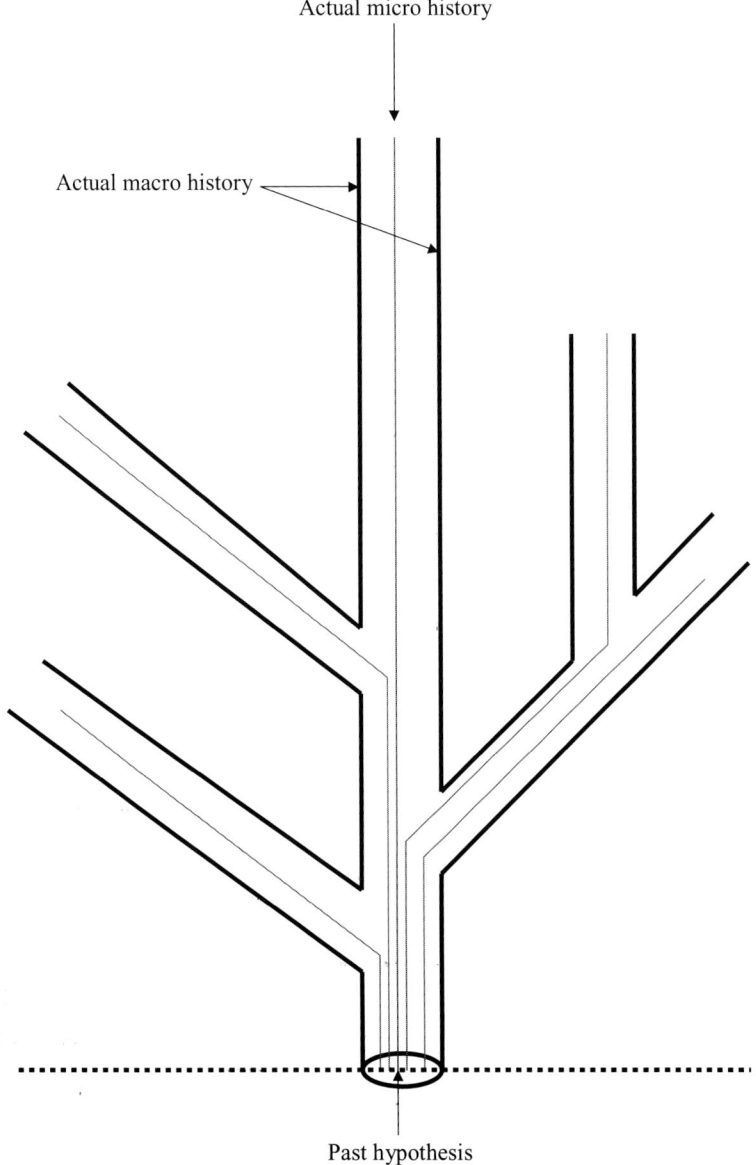

Figure 2.1.

possible macro-futures that get assigned relatively high probabilities conditional on the present macro-state. This contrasts with the probabilities assigned to different past evolutions:

From a typical macro state in the middle of the actual macro history there will be branching in both temporal directions but there will be much more branching where the branches have substantial probability in the direction away from the time of the *PH* than back towards it. The overall structure is due to the fact that the macro state at $t$ (in the middle) must end up in the direction of the boundary condition at which *PH* obtains (the direction we call 'the past') satisfying *PH*.

*(Loewer, 2007: 302)*

Loewer reiterates the same point further down:

[*PH* and *PROB*] determine an objective probability distribution over all nomologically possible micro histories (and *a fortiori* over all macro histories and all macro propositions). Even though the underlying micro dynamics is deterministic macro histories form a tree structure branching towards the future (away from the time at which *PH* holds).

*(Loewer, 2007: 307)*

That is, the objective probability distribution determined by *PH* and *PROB* forms a branching tree structure – a tree structure that is 'due to the fact' that *PH* provides a constraint on possible evolutions.

We can express this structure somewhat more formally by introducing the notion of quasi-determinism:

(QD) A system is *quasi-deterministic* at $t$ relative to some time $t'$ and some set of mutually exclusive macro-states $M$, exactly if there is a state $M_i$ in $M$ such that $P(M_i(t')/S(t))$ is close to 1, where $S$ is the state of the system at $t$.

The probabilities here (and throughout this chapter) are those induced by the statistical mechanical probability distribution and conditionalization on the dynamical micro-laws and the past hypothesis *PH* is left implicit. The claim that the universe exhibits an asymmetric tree-structure is equivalent to the conjunction of the following two claims:[4]

(1) the world is not quasi-future deterministic;
(2) the world is quasi-past-deterministic at all times $t$ with respect to all times $t'$, $t' < t$, and all $M$.

---

[4] Since Loewer only claims that there is 'much more branching' toward the future, one might think that conditions (1) and (2) are stronger than what Loewer would want to endorse. We do not need to settle this question in general here. The important issue will be whether there is widespread branching toward the past involving the kinds of event that feature in paradigmatically causal judgments.

I have shown in Frisch (2006a) that from a formalization of the tree-structure similar to QD one can indeed derive the asymmetry of decision counterfactuals that is at the core of Loewer's account. I have also suggested, however, that the claim that the SM account implies a tree-structure of possible macro-evolutions is problematic and here I want to develop this latter claim in more detail.

Since macro-states closer to equilibrium occupy vastly larger regions of phase space than states further away from equilibrium, it follows from Liouville's theorem that there will be many possible different non-equilibrium states far from equilibrium that evolve into the same state closer to equilibrium in the future. This suggests that there may be many more changes to the micro-state of a system close to equilibrium associated with different *macro-pasts* further away from equilibrium than there are changes to the micro-state of a system far from equilibrium associated with different *macro-futures* closer to equilibrium. Merely comparing the phase-space volumes associated with macro-states at different times suggests that possible macro-evolutions may exhibit an upside-down tree structure.[5]

We can distinguish two worries here: first, focusing on the future 'end' of the tree-structure, is it indeed a consequence of the SM account that there will be no significant reconvergence of branches, and that there are no times with respect to which thermodynamic systems are quasi-deterministic? And, second, focusing on the past 'end', is it a consequence of the account that the past is quasi-deterministic at all times with respect to the initial time $t_{PH}$ at which PH holds?

That future evolutions are not quasi-deterministic might seem to follow from the assumption of mixing. If a system is mixing the conditional probability $P(M(t)/M_0)$ of a macro-state $M(t)$ given the initial state $M_0$ depends only on the phase-space volume associated with $M(t)$ and is independent of $M_0$. Yet the mixing assumption alone does not imply the failure of quasi-determinism for all future times. If there is a single equilibrium macro-state $M_e$ that takes up the overwhelming majority of the phase space region available to a system, then $P(M_e)$ can be close to one and the system is quasi-future-deterministic with respect to all times after which the system reaches equilibrium.

This point holds for thermodynamic systems of all sizes – to the extent that the Boltzmannian account applies to these systems – ranging from small macroscopic quasi-isolated systems to the universe as a whole. Consider, for example, *the paradigmatic thermodynamic system* – a body of gas that is initially confined to

---

[5] Since Loewer represents possible macro-evolutions in his diagram of a possible tree structure (2.1) by cylinders of constant diameter, the diameter cannot be taken to represent phase space volumes. If we wanted to include a representation of the phase space volumes associated with macro-states in the diagram, possible macro-evolutions would have to be represented by cones of dramatically increasing widths towards the future. As a rhetorical device, Loewer's diagram lends far more plausibility to the thesis of macro-branching toward the future, than a picture of cones that branch at the same time as they dramatically increase in width. (When you try to draw this, you will quickly run out of space into which branching could occur.)

the right half of a container and, after a partition is removed, spreads out until it is distributed evenly throughout the container. Since most of the phase space accessible to the gas is associated with its equilibrium state, the SM account allows us to predict that the gas will be overwhelmingly likely to end up in that state – the gas evolves quasi-deterministically with respect to the final equilibrium state. At the other extreme, current cosmology suggests that the universe as a whole, too, may be quasi-deterministic with respect to its future equilibrium state, in which ionized stable particles, i.e. protons, neutrons and electrons, are distributed evenly throughout the cosmos at a density approaching zero (see e.g. Baez, 2007).

While Loewer is obviously right in suggesting that there are many systems that are open to the future – there clearly is widespread macro-branching toward the future – thermodynamic considerations imply that there also is widespread reconvergence of possible macro-histories. Thus, at the cosmological level, even though the initial state of the universe might not determine the large-scale distribution of matter before elementary particles begin to 'boil off' in the final evolution towards equilibrium, different cosmological macro-histories will converge toward the final equilibrium state. Similarly, there is convergence at the level of 'human-sized' macro-systems: no matter which part of the container a body of gas occupies initially, after the partition is removed the gas will spread until it is uniformly distributed throughout the container.

As a simple case exhibiting both branching and reconvergence consider the example Albert uses to motivate postulating a past hypothesis and the existence of macro-branching (Albert, 2000: 82ff). Albert imagines a system consisting of ice cubes that drop into glasses of water after sliding down a device similar to a Galton board. The same low-entropy initial state, with the ice cubes collected at the top, will indeterministically evolve into different macro-states given by different distributions of ice cubes in the glasses at the bottom of the board. Yet if we imagine that the ice cubes have several macroscopically distinct shapes of the same volumes, then there will be macroscopically distinct distributions of ice cubes in the glasses that will eventually evolve into the same macro-states once the ice is fully melted. And if we further assume that at the end of our experiment all glasses of water are emptied into a single bucket, all possible macro-histories that diverge after the ice is released at the top of the Galton board will reconverge – no matter what the shapes or volumes of the ice cubes are and no matter what path they take down the board. That is, the final state of the system when all the water is collected in a bucket is not quasi-deterministic with respect to past times when the ice cubes were distributed among the different glasses, even if we impose as additional constraint that all macro-histories are constrained to have originated in the state where the ice cubes were collected in a container at the top of the Galton board.

Now, there are discussions in the literature on counterfactuals that suggest that there is a crucial consideration that has been missing from our examination so far – the role of records or traces of the past. These discussions often invoke Kit Fine's famous example of Nixon's pushing the button that leads to a nuclear holocaust. It is often suggested that the many traces Nixon's action (or inaction) leave in the world play an important role in making convergence of 'button-pushing worlds' with 'non-button-pushing worlds' difficult. In the case of the ice cubes sliding down the Galton board drops of water on the board or my memories of observing a particular ice cube slide down a certain path might constitute such traces.

But it is easy to exaggerate how frequent and persistent macro-traces are. In fact, it is precisely the thermodynamic behaviour of systems that often either prevents the formation of macro-traces or leads to the disappearance of such traces. Whatever else the connection between *PH* and the existence of records is, one central role played by the thermodynamic arrow is as a destroyer of macro-records and macro-traces. Thus, any drops of water left on the Galton board will eventually evaporate; and since what path a particular ice cube took has not the same momentous consequences for the fate of the Earth as Nixon's decision whether to push the button, I will soon forget any details of what I observed (and may well forget altogether that I ever conducted the experiment). Moreover, there will not be any other macroscopic 'traces' of the experiment. While light waves will be reflected differently by the ice cubes depending on their path, due to the multiple scatterings of photons off the laboratory walls and the air molecules, these differences will leave no macroscopic traces by the next day. We might even imagine that there are different lamps that light up depending on what path an ice cube slides down. By the next day – and in fact much sooner – there will be no macroscopic traces of a particular lamp having been on when ice cube 17 slid down the board. Due to the thermodynamic behaviour of the walls of the laboratory and of the atmosphere, the macro-state of the world tomorrow will be independent of what the outcome of my experiment is today.

Examples similar to this one can be multiplied indefinitely. While there indubitably are many systems which for some finite time do not evolve quasi-deterministically, there are also many cases like the ones I just described – cases for which differences even in the current macro-state will eventually 'wash out', for which the system's macro-history throughout some period $T$ will leave no macroscopic traces in the future, and for which different macro-states will evolve quasi-deterministically into one and the same future macro-state.

I have argued that the assumption of mixing is not enough to ensure that a system is future quasi-indeterministic with respect to its equilibrium state and that it is a consequence of the thermodynamic behaviour of systems that there will be reconvergence of possible macro-histories even for systems that do not evolve quasi-deterministically during some time-interval $T$. Can mixing at least ensure that the evolution *toward* equilibrium is not deterministic? It is far from clear that the answer is 'yes'. First, all we can conclude from the assumption that a system is mixing is that after a sufficiently long time the probability of finding a system in a given macro-state is proportional to the phase-space volume associated with that state. That is, we can conclude from the fact that a system is mixing that it will *end up* in an equilibrium state, but we cannot draw any inferences at all about *how it will get there*.

Second, as Earman (2006) has argued, if we were able to show that all thermodynamic macroscopic systems had different possible macro-futures that receive substantial probability, we might be showing too much, as it were, and our theory would be empirically inadequate. While there clearly are systems that are dynamically unstable on the macro-level, there also are many systems that do not exhibit any macroscopic instability and are quasi-future-deterministic. Indeed, many paradigm cases of causal or time-asymmetric counterfactual judgments concern such quasi-deterministic macro-systems. Not only might we want to endorse the claim that had the proverbial butterfly not flapped its wings there would not have been a storm – an example of a causal counterfactual concerning a dynamically unstable system – but we might also want to say that had I not stepped on the brake, my car would not have come to a halt at the red light – an example of a causal counterfactual concerning, hopefully, a quasi-deterministic system. One might worry, then, how we can recover the apparently deterministic macro-evolutions of many systems from the assumption of dynamic instability on the micro-level.

The picture that has emerged is not one of an asymmetrically branching tree structure, but rather that of a web of possible macro-histories that branch and reconverge. Whether at its future end the web of possible macro-histories for the universe converges into a single strand, is a question for cosmology to decide. But the subweb characterizing the history of Earth and many of the even lower-dimensional 'subwebs' characterizing human-scale subsystems on Earth involve a large amount of convergence of strands, as well as branchings. Moreover, there are many small macro-systems that over (humanly) significant stretches of time evolve quasi-deterministically and do not exhibit any branching at all.

So far I have focused on Loewer's claim that there is branching toward the future without widespread reconvergence. I now want to turn to his claim that it follows from the SM-account that the macro-evolution of the universe is quasi-past-deterministic with respect to an initial time $t_{PH}$. Above I expressed the past hypothesis as the constraint that the initial macro-state of the universe was *a* low-entropy state satisfying certain further symmetry conditions. But if this indeed is what the past hypothesis says, an additional problem arises for Loewer's claim that the past is closed: It does not seem to follow from the constraint that micro-histories originated in *a* very low-entropy state that the macro-past is *the* unique actual low-entropy past. That is, counterfactual micro-histories may have originated in low-entropy macro-states *distinct* from the actual low-entropy past. Consider once more a system consisting of a gas in a box and assume that the gas could have started out in one of two possible low-entropy initial states, confined either to the right or the left half of the container. Let us assume that in the actual world the gas started out in the left half of the container and then spread out until it reached equilibrium, filling the entire container. Then, according to the reversibility objection, most changes to the final micro-state will be associated with a high-entropy past, since most micro-states compatible with the final equilibrium state will have evolved from equilibrium initial states. What if we assume a 'past-hypothesis' and constrain changes to the final micro-state to those that evolved from *a* low-entropy initial state? The phase space regions corresponding to the two initial states – the gas confined to the right or to the left half of the box – will evolve into highly fibrillated regions. If we assume that the system is mixing, each coarse-grained 'box' of phase space will have the same proportion of points that have evolved from the two initial regions. That is, intuitively, while the overwhelming majority of points in each box of phase space lie on trajectories that have evolved from high-entropy pasts, the same number of points in each box lies on trajectories that originated in the two low-entropy states. Given the final macro-state, the system is as likely to have evolved from the non-actual low-entropy past where the gas would have been confined to the right half of the container as from the actual past and adding a low-entropy constraint in the past does nothing to privilege the actual low-entropy past.

If *PH* merely restricts macro-histories to have originated in some (suitably symmetric) low-entropy state, then Loewer's conclusion that the universe is quasi-deterministic with respect to $t_{PH}$ seems unwarranted. But Loewer himself characterizes *PH* differently: He says that *PH* is 'a statement specifying *the* macro state of the universe at one boundary' (Loewer, 2007: 300, my italics). That is, according to Loewer's reading, the past-hypothesis restricts possible micro-histories to have originated in *the actual* low-entropy past state and this restriction trivially ensures that all possible macro-histories originated in one and the same macro-state. But can we assume the *actual* initial macro-state as constraint, without begging the question, in an account that is meant to derive a temporal asymmetry of counterfactuals?

Loewer's explicit aim is to provide a broadly Lewisian account of a counterfactual asymmetry and he contrasts his and Lewis's strategy, on the one hand, with Jonathan Bennett's, on the other, who does not offer an explanation of the asymmetry but simply assumes that counterfactuals are evaluated by keeping the past fixed. Loewer says:

I think that Bennett's account does a pretty good job of characterizing a conditional that matches core uses of the counterfactuals that interest us. [ ... ] However, Bennett's procedure for evaluating counterfactuals *assumes* the distinction between past and future (since forks are to the future) and so it does not provide a scientific explanation of time's arrows.

(Loewer, 2007: 309–310)

Thus, in order to provide a scientific explanation of the asymmetry, we cannot merely assume the asymmetry by holding the past fixed and allowing only the future to vary, but have to derive this asymmetry from the global distribution of matters of fact in the actual world in conjunction with the laws.

One might worry, then, that by the very fact that Loewer assumes *PH* as a time-asymmetric constraint he, like Bennett, is putting in the asymmetry by hand. Both Bennett and Loewer, it seems, stipulate a time-asymmetric constraint on how past states of the world may vary, and from this derive that counterfactuals are time-asymmetric. To be sure, Bennett's constraint goes beyond Loewer's – he stipulates that we hold fixed the *entire* macro-history in one temporal direction, while Loewer only fixes the macro-state at the past temporal *end* – but Loewer's constraint may strike one as similarly question-begging, if our goal is to provide a scientific explanation of a temporal asymmetry of counterfactuals. The only 'scientific contribution' to Loewer's account might be that the dynamical laws need to ensure that counterfactual past micro-evolutions converge quickly enough with the actual macro-past.

Loewer's reply to this worry is that the initial macro-state of the universe plays a special role in our overall scientific conception of the world. Loewer himself tries to capture this role by proposing a broadly Lewisian account of laws and suggesting that the actual initial macro-state is part of the Lewisian best system. Yet he apparently also believes that the special scientific status of the initial macro-state can be motivated independently of Lewis's account of laws. What, then, is the special role played by the *PH* and does that role provide us with good reasons for assuming the *actual* initial macro-state (rather than just *a* low-entropy state) as constraint on possible macro-histories?

First, in the Boltzmannian account, *PH* plays a central role in deriving the thermodynamic asymmetry. Thus, Loewer supports affording *PH* a special role by saying that it 'underwrite[s] many of the asymmetric generalizations of the special sciences especially those in thermodynamics and these generalizations are considered to be laws' (Loewer, 2007: 304). But in order to derive the thermodynamic 'laws' it is sufficient to assume that the universe began its life in *a* low-entropy state (in addition to *PROB*). Thus, the foundations of thermodynamics do not provide us with a reason to accept Loewer's version of *PH* as constraint instead of the one I proposed.

Second, both Albert and Loewer point to the explanatory role the actual macro-state of the early universe plays in current cosmology. Thus, they maintain that any macro-state that results from a small hypothetical alteration to the actual present macro-state is constrained to have evolved from the actual initial macro-state that cosmology will eventually present to us. But the plausibility of this claim relies on an equivocation on the notion of macro-state. A macro-state is associated with a coarse graining over the phase space of a system and in different contexts different coarse grainings are appropriate. In the case of the kind of counterfactuals associated with paradigmatically causal claims, the right level of description is one referring to medium-sized, 'human-scale' objects, whose states are characterized in units such as 1m or 1kg. In the context of astronomy or cosmology we are interested in the distribution of stars and galaxies and demanding that macro-states ought be specified to a precision of the location of medium-sized objects would be absurd. Appropriate units in the latter context are, for example, the astronomical unit $1\text{ AU} = 1.5 \times 10^{11}$ m or the solar mass $1\text{ M} = 1.9 \times 10^{30}$ kg. Thus, even if we grant that a specification of the actual initial macro-state of the universe provides a scientifically legitimate and non-question-begging constraint on possible macro-evolutions, the constraint can only be a specification of the *cosmological*, coarse-grained macro-state. Any specification of the initial macro-state more fine grained than that does not play a scientifically explanatory role.[6] But just as there are many *micro*-states compatible with a given fine-grained macro-state, there are many fine-grained macro-states (specifying, for example, the exact distribution of small rocks on a planet's surface) compatible with a more coarse-grained macro-state.

One might think that the specific nature of the macro-state of the early universe provides a reply to this worry. According to current cosmology, matter was distributed smoothly shortly after the big bang. (A smooth matter distribution, it is often argued, represents a state of extremely low gravitational entropy, and hence, as matter clumped to form stars and galaxies, the gravitational entropy of the universe increased.) Thus, one might think that there is just a single initial macro-state *tout court* – that is, even just a single fine-grained macro-state – that satisfies the conditions revealed to us by cosmology. While there can be many different macro-states that exhibit the same amount of gravitational clumping, there seems to be only a single macro-state characterized by a completely smooth matter distribution – a state that is smooth at all levels of coarse graining.

But this reply fails for two reasons. First, its premise is false. The macro-state of the early universe was not completely smooth, even on a cosmological level

---

[6] Within Loewer's preferred Lewisian account of laws this point can be made as follows: including a description of the Universe's fine-grained, human-scale initial macro-state in our deductive system will vastly complicate the system without providing us much (if any) gain in informativeness.

and – fortunately for contemporary cosmology – exhibited density fluctuations large enough to function as seeds for the formation of stars and galaxies.[7] Second, the inference from a distribution that is smooth at one level of coarse graining to one that is smooth at all levels is not sound. It is part of the Boltzmannian account that the micro-state of the early universe was one that is 'typical' given the known macroscopic constraints. This means that, if the association between a smooth matter distribution and low gravitational entropy is correct, the SM account implies that the early universe is overwhelmingly probable to have exhibited as much gravitational clumping as is compatible with our cosmological evidence.

Thus, neither statistical physics nor cosmology provides us with scientific reasons to take the actual *fine-grained* or human-scale macro-state as constraint on possible fine-grained macro-evolutions. The Boltzmannian account requires as premise only that the universe began its life in *a* state of extremely low entropy and cosmology restricts that state to one that is characterized by an approximately smooth matter distribution, but with density fluctuations large enough to be compatible with many different fine-grained macro-states.

## 2.4 Causal handles

In the previous section I argued that it is a consequence of the thermodynamic arrow that there will be convergence among different possible macro-evolutions and that there will be many cases where small differences in the macro-state of a system at one time leave no macro-traces in the system's future. In this section and the next I will show that this result leads to problems both for Albert's account of causal handles and for Loewer's account of decision counterfactuals.

Albert argues that it is a consequence of the Boltzmannian account that the present contains multiple *causal handles* on the future but (almost) no causal handles on the past. If we constrain the *remote* past of any physical system, he maintains, then only very few and special alterations to the present are associated with a different *recent* past, while many such alterations may lead to different futures. To illustrate this point Albert asks us to consider a collection of idealized billiard balls on a frictionless plane such that ball 5 is currently stationary with the additional constraint that ball 5 was moving 10 seconds ago. Given this additional constraint, the fact that ball 5 has been involved in a collision in the past 10 seconds is nomically *determined* by facts about the present state of ball 5 *alone*. That is, alterations to the present state of the balls *not* involving changes in the state of ball 5 cannot change the fact that ball 5 was involved in a collision during the last

---

[7] The density fluctuations are of the order of 1 part in 100 000. By comparison, differences in mass distribution of interest to us are of the order of $10^{-30}$ times the mass of the sun.

10 seconds. Yet there are many changes to the state of the balls not involving ball 5 that will result in a different future evolution of ball 5. From this Albert concludes that there are a far wider variety of 'what we might call *causal handles* on the future of the ball in question here, under these circumstances, than there are on its past' (Albert, 2000: 128). In this example the constraint that ball 5 was moving is meant to play the role of a 'past-hypothesis' and the current state of ball 5 functions as a record of the past collision. More generally, then, Albert claims that if we postulate *PH* as constraint on all possible macro-evolutions, then this imposes almost no additional restriction on possible future macro-evolutions, while it restricts non-actual present macro-states that are the result of small macro-changes to the actual present state to have evolved from the *actual* macro-past – that is, it follows from imposing *PH* as constraint on all possible macro-histories that there are many more causal handles on the future than on the past.

In Albert's example the current state of ball 5 together with the past constraint nomologically determine that ball 5 was involved in a collision. In the general case, however, *PH* in conjunction with certain local facts about the current macro-state assigns probabilities strictly less than one to the occurrence of past events. Many records or traces of the past do not determine the occurrence of the events of which they are records but only raise the probabilities of their occurrence. Thus, the general definition of a causal handle is as follows: A macro-event $C(t)$ is a *causal handle* on an event $E(t')$, just in case the occurrence of $C$ (significantly) affects the probability of $E$. That is, $C(t)$ is a causal handle on $E(t')$ exactly if $P(E/C\&M(t)) \neq P(E/M)$. $M(t)$ is the actual macro-state at $t$ outside of the region where $C$ occurs and contains any putative records of $E$ at $t$. On Albert's proposal, we evaluate the results of small hypothetical changes to the present by keeping the present macro-state fixed except for the small change and then determine how this counterfactual macro-state evolves in accord with the constraint given by the SM account – with one important qualification: Albert assumes that, in addition to any macro-records of an event, we also hold fixed any putative *memories* we might have of that event, even though memories might be physically realized by micro-states.

Albert's thesis that there are (almost) no causal handles on the past is tantamount to a screening-off condition. $C(t)$ is not a causal handle on some past event $E$ just in case the rest of the macro-state at $t$ screens off $E$ from $C$ – that is, $P(E/C\&M(t)) = P(E/M)$. But for events $E$ that leave at most a small number of traces in the present, this condition can easily fail. Take an event $E(t')$ that has only two distinct macro-traces $C_1$ and $C_2$ at some later time $t$. $C_1$ and $C_2$, intuitively, are both effects of $E$. Thus, while $E$ as the common cause of $C_1$ and $C_2$ might screen off $C_1$ from $C_2$, it will not in general be the case that Albert's condition is satisfied and that one effect screens off the cause from the other effect. Indeed, the presence (or absence) of additional traces of an event – that is of additional evidence for the event's occurrence – can radically alter the probability of that event. Albert's Galton board can again serve as an example. That a particular ice cube landed in the second glass from the left, say, constitutes a trace of it having slid down to the left of the first pin. (For a board with $n$ rows of pins where the probability at each pin of the ice cube sliding down on one side is equal to 1/2, this probability is $P$(left/second glass to left) $= 1 - 1/n$.) Now let us imagine that the ice cube can dislodge a little ball as it slides down the board and that where the ball ends up also functions as a probabilistic record of the ice cube's path down the board. It is then easy to set up the probabilities in such a way that both the little ball's present position and the ice cube's landing in the second glass come out as causal handles on the path of the ice cube past the first pin. That is, it is easy to set things up such that (keeping the present condition of the ice cube in the glass fixed) there can be many alterations in the present condition of the little ball that would alter the probabilities about whether or not the ice cube slid down to the left of the first pin.[8]

Again, there is nothing unusual about this example. There are many cases where additional evidence affects the probabilities of past events, and hence, according to Albert's account, would classify as causal handle on the past. It seems to me that to the extent that Albert's thesis may appear intuitively plausible, this rests on at least one of the following two assumptions. First, Albert's thesis is true if we demand that a trace (together with the past-hypothesis) nomologically *determines* the event's occurrence. But while this assumption might hold in the idealized billiard ball case Albert considers, it clearly is not true in general. Secondly, the thesis would follow, if we assumed that each event leaves sufficiently many and varied traces that each trace taken individually only marginally affects the probability of the event's occurrence. But as I argued in the previous section, it is a consequence of the thermodynamic arrow that this assumption is often false. There are many mundane (and paradigmatically causally related) events that leave no or only very few traces in the future.[9]

## 2.5 Decision counterfactuals

By contrast with Albert's account, Loewer's account of an asymmetry of control focuses primarily not on a purely physical asymmetry but on an asymmetry involving agents. Loewer argues that the SM account underwrites an asymmetry of *decision counterfactuals*. A decision counterfactual is a probabilistic counterfactual of the form 'If at $t$ I were to decide $D$, then the probability of $B$ would be $P$',

---

[8] And this is meant as an explicit contrast with what Albert says about the billiard ball case on the top of page 127 in his book.
[9] I critically examine several other aspects of Albert's account of the causal asymmetry in Frisch (2007).

which is true exactly if $P(B/M(t)\&D(t)) = p$. $M(t)$ is the complete macro-state at $t$ and the decision $D(t)$ is an event 'smaller than a macro-event but with positive probability' (Loewer, 2007: 316). A property of decision events that is attractive from Loewer's perspective is that small differences in a decision state can get magnified into large macroscopic differences in the world and he maintains that this feature of 'decision conditionals [is] temporally asymmetric': 'Alternative decisions that can be made at time $t$ typically can make a big difference to the probabilities of events after $t$ [...] but make no difference to the probabilities of macro events prior to $t$' (Loewer, 2007: 317).

Trying to capture the idea that different decisions are 'open' to an agent at a time, Loewer assumes that decisions are 'indeterministic relative to the macro state of the brain and environment prior to, and at the moment of making the decision' (Loewer, 2007: 317). From this assumption, it seems, there is an extremely quick argument for the asymmetry of decision counterfactuals. If the assumption is understood not merely as denying determinism but as asserting that decisions are probabilistically independent of the macro-state prior to $t$, it directly follows that differences in decisions 'make no difference to the probabilities of macro-events prior to $t$'. But this argument for the asymmetry of decision counterfactuals does not rely on the SM account at all and seems question-begging – the asymmetry of decision counterfactuals is simply built into our account of what a decision is. If we want to avoid begging the question, we need to treat Loewer's decision counterfactuals analogous to Albert's causal handles: in evaluating the truth of counterfactual we hold the actual present *macro*-state and, in addition, our present memories fixed, posit an alternative decision-event compatible with the state we keep fixed and then let the conditional probabilities of both future and past macro-events be those given by the SM account. For the account to succeed, Loewer's thesis that alternative decisions make no difference to the probabilities of past events would have to come out as a consequence of the SM account.

Yet the SM account fails to imply Loewer's thesis. As a matter of fact our decisions at $t$ are not completely independent of the macro-state of the world prior to $t$ – many of my decisions today reflect facts about my biography and are strongly correlated with past experiences. While there may be decisions which amount to mere random 'picking' and hence may be probabilistically independent of my past, many of my decisions exhibit a certain coherence and represent facts both about who I am and about the world.[10] That is, for many of my decisions there are events $B$ in the past such that $P(B/D(t)) \neq P(B/\text{not-}D(t))$. Moreover, acknowledging this dependence does not force us to deny Loewer's assumption that different decisions or choices are 'open' to an agent making a decision, since plausibly, this assumption

---

[10] For a discussion of the distinction between 'picking' and 'choosing' see Holton (2006).

can be captured by supposing that an agent's beliefs and desires do not determine her choices (see Holton, 2006: 4) and this supposition is compatible with the claim that an agent's choices are probabilistically correlated with events in her past. Finally, even though my history plays a role in shaping the choices I make, I consciously remember only very few events of my past and only very few of these events have left completely reliable macroscopic traces in the present. Thus, the present macro-state in conjunction with my memories does not screen off the past from my decisions – that is, for many of my decisions it will be the case that there are past events $B$ such that $P(B/M(t)\&D(t)) \neq P(B/M(t))$.[11]

Thus, many of an agent's decisions do make a difference to the probabilities of macro-events prior to the time of her making the decision. Now, Loewer argues that even in these cases there still is an important asymmetry between past and future correlations, since, he maintains, we cannot have *control* over events in our past. According to Loewer, the condition of having control is strictly stronger than the condition of probabilistic dependence: 'control by decision requires that there be a probabilistic correlation between the event of deciding that p be so and p being so and one's knowing (or believing with reason) that the correlation obtains' (Loewer, 2007: 318). Loewer's first condition on control is that there has to be a probabilistic correlation between a decision $D$ and the event $B$ over which we have control. This condition, I have argued, is satisfied for large sets of pairs of decisions and events in their past.

A second condition is that we must have good reasons to believe that such a correlation obtains. This condition, too, appears to be frequently satisfied, since we are often in a position to discover how our decisions are correlated with our history. Holton argues that one important role for decision or choice in our lives might be that it enables agents to come to know something about themselves *and* about the world that they would not have been in a position to know prior to their decision (Holton, 2006). According to Holton, by looking at their choices, agents 'can form, rather than just discover, their judgments on that basis' (Holton, 2006: 10–11). On Holton's account, an agent who has the right kinds of competences can in certain circumstances learn from her decisions, since her decisions can act like a finely tuned instrument that picks up on cues that are not consciously available to the agent. If some account like this is correct, then there are correlations between our decisions and past events that we can come to believe with reason and Loewer's second condition is satisfied as well for certain past events.

---

[11] As a putative counterexample to his account Loewer asks whether my decisions now can affect the existence of Atlantis. One implication of my discussion here is that this is not the kind of counterexample about which Loewer should worry. Much more worrisome than the case of Atlantis for Loewer's account are events in the past that the agent facing a decision experienced, but that left (almost) no traces in the present.

Yet one might think that while there are many cases where one of the two conditions is satisfied individually, the two conditions can never be jointly satisfied when the events in question lie in the past of a decision. One might think that we can learn of correlations between our decisions and past events only when we remember these events or are in the possession of other reliable records of them, but to the extent that our memories or records are reliable they screen off the past experiences from our present decisions. That is, when the second condition is satisfied, the first condition fails. By contrast, when $P(B/M(t)\&D(t)) \neq P(B/M(t))$, we cannot rely on any records to come to know the correlation between our decisions and a past event $B$. That is, when the first condition is satisfied, the second condition fails.

But this objection can be answered. We can learn of correlations between certain kinds of decisions and past events when we do have memories of the past events in question and then use that knowledge inductively to learn something about the past in cases where we make similar decisions but where the relevant memories are absent. This is not much different from how we come to believe reasonably that our decisions are correlated with future events – by learning inductively from experiences of past correlations.

Here is an example of this. While playing a piano piece that I know well I am unsure whether I am currently playing a part of the piece that is repeated in the score for the first or the second time. I decide to play the second ending rather than repeating the part. Many of the notes I play, of course, I play without choosing or deciding to play them. But in the case I am imagining the question what notes to play next has arisen, and I consciously choose to play the second ending. Since I have learned from experience that when I play a piece I know well, my decisions to play certain notes are good evidence for where I am in the piece, my present decision not to repeat the part constitutes good evidence for a certain past event – my having already played the part in question once. We can even imagine that I have a vague and unreliable memory of having already played the part. My decision to play the second ending then can constitute additional evidence for the reliability of my memory. In general, Loewer's first and second conditions are jointly satisfied in cases in which (i) we have good (inductive) reasons for treating our decisions as providing us with information about our past or past events in the world and (ii) the past events in question have left no or only very few and not fully reliable traces in the present.

As a third condition Loewer requires that we have control over an event $B$ only if it is part of the content of our decision that $B$ occur. This third condition is not satisfied for events $B$ in the past of the decision in the case of past events, since we do not (normally) take ourselves to have control over the past. But I think this last condition is too strong. We take ourselves to have control over events that are consequences of our decision, even when the content of our decision is not that these events occur. For example, I may have the desire to arrive at the office by 9 a.m. and I have good reasons to believe that my arrival time is reliably correlated with the time when I leave my home. Then my decision to leave at a certain time provides me with a means of controlling when I arrive, even though my decision is, say, a decision to leave at 8 a.m. rather than a decision to arrive by 9 a.m. It seems that we can have control that $p$ be so, by decision, even when our decision is not a decision that $p$. Thus, two of Loewer's conditions on control by decision can be jointly satisfied by past events, while the third condition should be rejected on independent grounds.

## 2.6 Conclusion

I have argued that, contrary to what Loewer suggests, it is a consequence of the SM account that there are reconvergences and mergings of possible macro-histories, as well as branchings. Moreover, the thermodynamic asymmetry results in the destruction of records or traces of the past and there are many 'human-scale' macro-events that leave no or only very few traces in their futures. This result presents serious problems both for Albert's and Loewer's accounts of the asymmetry of causal influence and control. If what I argued in this chapter is correct and if the SM account allows that small alterations to the present macro-state and changes to our decisions are probabilistically correlated with changes both to the macro-future and to the macro-past, then Albert and Loewer still owe us an account of how we came to posses strictly time-asymmetric concepts of causal influence or control.

## References

Albert, D. Z. (2000). *Time and Chance*. Cambridge, MA: Harvard University Press.
Baez, J. (2007). *The End of the Universe* 2004 (accessed 1 February 2007). Available from: Causation in Physics/Baez_universe_end.html.
Earman, J. (2006). The 'Past Hypothesis': not even false. *Studies in the History and Philosophy of Modern Physics*, **37**(3), 399–430.
Frisch, M. (2000). (Dis-)solving the puzzle of the arrow of radiation. *British Journal for the Philosophy of Science*, **51**, 381–410.
Frisch, M. (2005). *Inconsistency, Asymmetry and Non-locality: a Philosophical Investigation of Classical Electrodynamics*. New York: Oxford University Press.
Frisch, M. (2006a). Causal asymmetry, counterfactual decisions and entropy. *Philosophy of Science*, **72**(5), 739–750.
Frisch, M. (2006b). A tale of two arrows. *Studies in the History and Philosophy of Modern Physics*, **37**(3), 542–558.
Frisch, M. (2007). Causation, counterfactuals and the past-hypothesis. In *Russell's Republic: the Place of Causation in the Constitution of Reality*, ed. H. Price and R. Corry. Oxford: Oxford University Press.
Holton, R. (2006). The act of choice. *Philosophers' Imprint*, **6**(3), 1–15.

Loewer, B. (2007). Counterfactuals and the second law. In *Causality, Physics, and the Constitution of Reality: Russell's Republic Revisited*, ed. H. Price and R. Corry. Oxford: Oxford University Press.

Uffink, J. (2006). Compendium to the foundations of classical statistical physics. In *Philosophy of Physics, Handbooks of the Philosophy of Science*, ed. J. Butterfield and J. Earman. Amsterdam: Elsevier, North-Holland.

Winsberg, E. (2004). Can conditioning on the 'past hypothesis' militate against the reversibility objection? *Philosophy of Science*, **71**, 489–504.

# 3

# The past hypothesis meets gravity

CRAIG CALLENDER

## 3.1 Introduction

Why does the universe have a thermodynamic arrow of time? The standard reasoning relies on the truism: no asymmetry in, no asymmetry out. If the fundamental laws of Nature are time symmetric invariant (that is, time reversal), then the origin of the thermodynamic asymmetry in time must lie in temporally asymmetric boundary conditions. However, this conclusion can follow even if the fundamental laws are not time reversal invariant. The more basic question is whether the fundamental laws – whether time symmetric or not – entail the existence of a thermodynamic arrow. If not, then the answer must lie in temporally asymmetric boundary conditions. No asymmetry of the right kind in, no asymmetry out. As it happens, as I understand them, none of the current candidates for fundamental law of Nature entail the thermodynamic arrow. String theory, canonical quantum gravity, quantum field theory, general relativity, and more all admit solutions lacking a thermodynamic arrow. So a first pass at an answer to our initial question is: the universe has a thermodynamic arrow due in part to its temporally asymmetric boundary conditions.

Merely locating the answer in boundary conditions, however, is not to say much. All it does is rule out thermodynamic phenomena being understood as a corollary of the fundamental laws. But that's true of almost all phenomena. Few events or regularities can be explained directly via the fundamental laws. If we are to have a satisfying explanation, we need to get much more specific.

One promising way of doing so is via the explanation Boltzmann initially devised. Roughly put, the idea is as follows. Identify the thermodynamic entropy of a system with the so-called Boltzmann entropy. Then make plausible the claim that if the initial Boltzmann entropy of the system is low, then over 'reasonable' time spans in the future it is highly likely that it will increase. Finally, assume as a boundary condition that the initial Boltzmann entropy of the system was low. With these pieces in place, one can infer that the system will display a thermodynamic arrow over the time spans in question. What is the system to which this applies? Because it is difficult to know how to decouple systems in a non-arbitrary way, Boltzmann took himself to be describing the entire universe. If this is right, we now have a theory explaining why we have a thermodynamic arrow in our universe.[1] And this explanation appeals to a much more specific claim about boundary conditions than the generic reasoning we engaged in above: namely, that the Boltzmann entropy of the entire universe was very low (compared to now) roughly 15 billion years ago. In particular, the entropy of this state was low enough to make subsequent entropy increase likely for many billions of years. Let us follow Albert (2000) in calling this claim the past hypothesis; let us call the state it posits the *past state*. Physicists such as Boltzmann, Einstein, Feynman, Penrose and Schrödinger have all posited the past state in one form or other. To me, if the Boltzmann framework can be defended, then positing the past state in one form or other appears to be the simplest answer to the problem of the direction of time in statistical mechanics.

Simplicity is nice, but truth is better. Is the past hypothesis true? When we look to the early universe, as described by contemporary cosmology, do we observe something resembling the past state? Some authors (e.g. Price, 1996) believe that Boltzmann's prediction is spectacularly well confirmed by cosmology. I agree that if correct, the vindication of Boltzmann's novel retrodiction should count among the great achievements of science. It would be a prediction of the early state of the universe from seemingly independent statistical mechanical arguments. However, if Boltzmann's prediction is right, why is it so unsung? The answer is that we cannot be confident that the prediction is right. The reason for this is that it has never been entirely clear how to apply Boltzmann's statistical mechanical framework in conditions such as those in the early universe.

Bracket all the questions still under debate about the big bang. Let us not worry about cosmic inflation periods, the baryogenesis that allegedly led to the dominance of matter over anti-matter, the spontaneous symmetry breaking that purportedly led to our forces, and so on. The past state does not have to be the 'first' moment. Skip to $10^{-11}$ seconds into the story when the physics is less speculative. Or skip even further into the future if you are worried about the standard model in particle physics. (And do not even think about dark energy or dark matter.)

Even still, for confirmation of Boltzmann's insight, at the very least one needs to understand statistical mechanics and Boltzmann entropy in generally relativistic spacetimes, the entropy of radiation, how this entropy relates to the entropy of

---

[1] Many physicists and some philosophers want *more*: they want to explain why the boundary condition is what it is. In Callender (2004) I argue that this is not necessary.

the matter fields, and more. Needless to say, all of this is highly non-trivial.[2] That the physics needed is largely unknown, of course, does not imply that the past hypothesis is false; it merely explains why Boltzmann is not lauded or faulted for his prediction.

What would suggest falsity is if – as a matter of principle – the basics of Boltzmann's framework just cannot be applied in the non-classical theories needed to describe the physics of the early universe. That is Earman's (2006) claim with respect to general relativity. In a sharp attack, Earman claims that the past hypothesis is 'not even false'. The reason for this conclusion is that Earman is unable to define a coherent and non-trivial Boltzmann entropy in general relativity. For the Boltzmann entropy to make sense (as we will see) one needs a well-defined state space for the theory and a measure invariant under dynamical evolution. We do not have this for the space of all solutions to Einstein's field equations. We do have it for some very special cases. Restricted to Friedman–Robertson–Walker metrics with a scalar matter field, one can use the Hawking–Page measure over a two-dimensional reduced phase space or the Holland–Wald measure over a three-dimensional reduced phase space. Earman shows that using either makes nonsense of the past hypothesis.

Earman's result is troubling, but perhaps not fatal to the past hypothesis.[3] The measures he cites, it must be admitted, are developed only for a highly idealized set of solutions to Einstein's field equations. The problem of developing measures on the space of solutions to Einstein's field equation is still in its infancy, e.g., we are very far from claiming that either of the above measures is uniquely invariant with respect to time evolution. Hence we do not have an in principle demonstration that the Boltzmann entropy is indefinable in general relativity. There may be other measures that work. What Earman shows is that given what we know, things do not look good.

Given this situation, a natural question is whether the Boltzmann entropy makes sense even in classical physics when we consider cosmological systems. In particular, since what is causing the present trouble is gravity, one would like to understand an early classical state when the gravitational interactions are included in the system. Such an approach would be deeply limited. As mentioned, to describe anything like our universe one needs general relativity, the expansion of space, strong and weak nuclear forces, and much more. While this claim is no doubt true, there are virtues in beginning simply. For if we have trouble even here, then we know we

---

[2] Please do not be fooled into thinking that various entropies used in the literature, including the so-called Bekenstein entropy, are a quick fix. It is commonly asserted that the Bekenstein entropy is low in the past and high in the future. Without a clear connection between this entropy and the Boltzmann entropy, however, this claim simply is not relevant to our question.

[3] Earman also launches other attacks on the past hypothesis and the uses to which it has been put, but we only have space to focus on this problem.

have a problem with gravity no matter how the measure-theoretic details work out in general relativity. And if some problems in the classical context can only be solved by adding non-classical elements, then that is still something interesting to learn. Before worrying about general relativistic or quantum gravitational thermodynamics, let us figure out whether classical gravitational thermodynamics works.

As we shall see, even here in the Newtonian context – surprisingly – matters get tremendously complex. Nasty 'paradoxes' threaten the very foundations of gravitational thermodynamics. The point of the present chapter is to introduce these problems and show how they affect the Boltzmann explanation described above.

This chapter has two very modest goals. Firstly, and primarily, I want to demonstrate why even classical gravity is a serious problem for the standard explanation of entropy increase. If the chapter does nothing else, my hope is that it gets the problems induced by gravity the attention they deserve in the foundations of physics. Secondly, I want to outline a possible way out of at least one difficulty. Most of the work here will be in the set-up, both in seeing the exact nature of the problem and in understanding how the work done on the statistical mechanics of stellar systems can be conceived from a foundational perspective. Once framed, I want to make plausible a very weak claim: that there is a well-defined Boltzmann entropy that *can* increase in *some* interesting self-gravitating systems – where I get to define 'interesting'. More work will need to be done to see if this claim really answers the threat to the standard explanation of entropy increase. However, establishing the claim might remove some of the pessimism one might have about the standard explanation in the gravitational context, in addition to suggesting a clear path for future study.

## 3.2 The past hypothesis

Classical phenomenological thermodynamics is a system of functional relationships among various macroscopic variables, e.g., volume, temperature, pressure. It tells us that some macro-states $M$ covary or evolve into others, e.g., $M_{t_1} \to M_{t_2}$. One of these relationships is the famous second law of thermodynamics. It tells us that an extensive state function S, the entropy, defined at equilibrium, is such that changes in it are either positive or zero, i.e. entropy does not decrease. For realistic cases, it seems to imply that in the spontaneous evolution of thermally closed systems, the entropy increases and attains its maximum value at equilibrium. Actually, there is controversy whether the spontaneous movement from non-equilibrium to equilibrium strictly follows from the second law; but even if it does not, there is no controversy that this spontaneous movement occurs and is a central feature of thermodynamics. This feature describes many of the temporally directed aspects of our world, e.g., heat going from hot to cold, gases spontaneously expanding throughout their available volumes.

Why, from a mechanical perspective, do these temporally directed generalizations hold? Let us restrict ourselves to classical statistical mechanics, and in particular, the Boltzmannian interpretation of statistical mechanics. I find the Boltzmannian view of statistical mechanics provides a more 'physical' description of what is going on from a foundational perspective than the rival Gibbsian perspective.[4]

The first step in understanding the Boltzmannian explanation of the approach to equilibrium is distinguishing the macroscopic from microscopic description of the system. The exact microscopic description of an unconstrained classical system of $n$ particles is given by a point $X \in \Gamma$, where $X = (\mathbf{q}_1, \mathbf{p}_1 \ldots \mathbf{q}_n, \mathbf{p}_n)$ and $\Gamma$ is a $6n$-dimensional abstract space spanned by the possible locations and momenta of each particle. $X$ evolves with time via Hamilton's equations of motion. Since energy is conserved, this evolution is restricted to a $6n - 1$ dimensional hypersurface of $\Gamma$.

The same system described by X can also be described in the macro-language by certain macroscopic variables (volume, pressure, temperature, etc.). This characterization picks out the system's macro-state $M$. Notice that many other micro-states will also give rise to the same macro-state $M$. If we consider all the $X \in \Gamma$ that give the same values for macroscopic variables as $M$ gives, this will pick out a volume $\Gamma_M$. The set of all such volumes partitions is the energy hypersurface of $\Gamma$.

A quick word about the volume. A continuous infinity of micro-states will give rise to any particular macro-state, so one requires the resources of measure theory. The $(6n-1)$-dimensional energy hypersurface of $\Gamma$ has a Lebesgue measure naturally associated with it. From this measure one creates a probability measure, and one assumes or hopes to prove that the probability of finding a system in region $\Gamma_M$ of the energy hypersurface of $\Gamma$ is proportional to the volume of $\Gamma_M$, $|\Gamma_M|$, within $|\Gamma|$.

We can now define the entropy of a macro-state M. The *Boltzmann entropy* of a system $X$ that realizes $M$ is defined by

$$S = k \log |\Gamma_M(X)|$$

where $k$ is Boltzmann's constant and $\| \|$ indicates volume with respect to Lebesgue measure. Notice that this entropy is defined in and out of equilibrium. In equilibrium, it will take the same value as the Gibbs fine-grained entropy if $n$ is large.

---

[4] I am not alone. Lavis (2005: 246) writes, 'When confronted with the question of what is "actually going on" in a gas of particles (say) when it is in equilibrium, or when it is coming to equilibrium, many physicists are quite prepared to desert the Gibbsian approach entirely and to embrace a Boltzmannian view'. See Lavis for a description of the Gibbsian view.

Outside equilibrium, the entropy can take different values and will exist so long as a well-defined macro-state exists.

Why should Boltzmann entropy increase? The answer to this is controversial, and we do not have space to discuss it fully here. The hope is that one will be able to show that *typical* micro-states underlying a non-equilibrium macro-state subsequently head for equilibrium. One way to understand this is as follows.[5] The Boltzmann equation describes the evolution of the distribution function $f(\mathbf{x}, \mathbf{v})$, over a certain span of time, and this evolution is one toward equilibrium. Let $\Gamma_\delta \subset \Gamma$ be the set of all particle configurations $X$ that have distance $\delta$, $\delta > 0$, from $f(\mathbf{x}, \mathbf{v})$. A *good* point $X \in \Gamma_\delta$ is one whose solution (a curve $t \to X(t)$) for some reasonable span of time stays close to the solution of the Boltzmann equation (a curve $t \to f_t(\mathbf{x}, \mathbf{v})$). A *bad* point $X \in \Gamma_\delta$ is one that departs from the solution to the Boltzmann equation. The claim that typical micro-states underlying a non-equilibrium macro-state subsequently head for equilibrium is the statement that, measure theoretically, most points $X \in \Gamma_\delta$ are good. The expectation – proven only in limited cases – is that the weight of good points grows as $n$ increases. The Boltzmannian wants to understand this as providing warrant for the belief that the micro-state underlying any non-equilibrium macro-state one observes is almost certainly one subsequently heading toward equilibrium. As mentioned, the desired conclusion does hang on highly non-trivial claims, in particular, the claim that the solution to Hamilton's equations of motion for typical points follows the solution to the Boltzmann equation.

Here is a loose bottom-to-top way of picturing matters that will come in handy later (see DeRoeck, Maes and Netočný, 2006). We know at the macroscopic level that non-equilibrium macro-states evolve over short periods of time into closer-to-equilibrium macro-states. That is, $M_1$ at $t_1$ will evolve by some time $t_2$ into a closer-to-equilibrium macro-state $M_2$. Call $\Gamma_{M_1 t_1}$ the set of states in $\Gamma$ corresponding to $M_1$ at $t_1$, $\Gamma_{M_2 t_2}$ the set corresponding to $M_2$ at $t_2$, and $\phi_{t_2 - t_1} \Gamma_{M_1 t_1}$ the time evolved image of the original set $M_1$. Then, if our picture is right, the second law is telling us that $\phi_{t_2 - t_1} \Gamma_{M_1 t_1}$ is virtually a proper subset of $\Gamma_{M_2 t_2}$. That is, almost all of the points originally in $M_1$ have evolved into the set corresponding to $M_2$. Liouville's theorem states that a set of points retains its size through Hamiltonian evolution. Hence the volume of $\phi_{t_2 - t_1} \Gamma_{M_1 t_1}$ is equal to the volume of $\Gamma_{M_1 t_1}$. Since the former is virtually a proper subset of $\Gamma_{M_2 t_2}$, that means that $|\Gamma_{M_1 t_1}| \leq |\Gamma_{M_2 t_2}|$. From the definition of entropy it follows that $S(M_{t_2}) \geq S(M_{t_1})$.

The problem of the direction of time is simple to see. Nowhere in the above argument did I say whether $t_2$ is before or after $t_1$. Given a non-equilibrium state at

---

[5] For a general discussion, see Goldstein (2002) and references therein. For the specific formulation here, see Spohn (1991: 151). And for some of the challenges this approach faces, see Frigg (2009).

$t_1$, the above reasoning shows that it is very likely that it will subsequently evolve to a later higher entropy state at $t_2$, where $t_1$ is earlier than $t_2$. However, it is also true that the reasoning shows that most likely the state at $t_1$ evolved from an earlier higher entropy state, in this case where $t_2$ is earlier than $t_1$. There is nothing in the time reversible dynamics nor in the above reasoning to rule out entropy increase in both temporal directions from the non-equilibrium present. The famous recurrence and reversibility challenges to Boltzmann point out that even good points $X$ will go bad if given enough time (recurrence) or allowed to go in the wrong temporal direction (reversibility).

All manner of answers to this problem have been proposed – appeals to time asymmetric environmental perturbations, ignorance, electromagnetism, and more. In my opinion, where these proposals have merit, they eventually reduce to an appeal to temporally asymmetric boundary conditions. Ultimately we need to assert that in the direction we call 'earlier' entropy was in fact very low compared to now. As mentioned at the outset, the specific form of this claim in the present context is that the past hypothesis is true; that is, that the Boltzmann entropy of the universe was extremely low roughly 15 billion years ago.

## 3.3 The past hypothesis meets gravity

No sooner is the past state posited than it is immediately challenged with a bit of a problem: it seems to be manifestly false. When we look to cosmology for information about the actual past state, we find early cosmological states that appear to be states of very high entropy, not very low entropy. Cosmology tells us that the early universe is an almost homogeneous isotropic state of approximately uniform temperature, i.e. a very high entropy state, not a low-entropy state as mandated by the past hypothesis. Here is the physicist Wald:

> The above claim that the entropy of the very early universe must have been extremely low might appear to blatantly contradict the 'standard model' of cosmology: there is overwhelmingly strong reason to believe that in the early universe matter was (very nearly) uniformly distributed and (very nearly) in thermal equilibrium at uniform temperature. Does not this correspond to a state of (very nearly) maximum entropy, not a state of low entropy?
>
> *(Wald, 2006: 395)*

If we consider point particles interacting without gravity, then the answer certainly seems to be in the affirmative.

Once the problem is stated, however, authors quickly reassure us that it is only apparent. We forgot to include gravity, we are told, and yet by including gravity the 'situation changes dramatically' (Wald, 2006: 395). Gravity saves the past hypothesis. This claim is made with equal frequency and force by scores of physicists and philosophers of physics.

How does gravity save the past hypothesis? Here is a (too) simple expression of the idea. If we think of a normal terrestrial gas in a box, as a result of repulsive forces and collisions, its 'natural tendency' is to spontaneously spread throughout its available volume into a homogeneous state. If this is right, then when we add an attractive force like gravity the reasoning should reverse. For it is the 'natural tendency' of a gravitating system to spontaneously move toward more clumped states. Masses attract one another, and both in theory and computer simulation self-gravitating Newtonian systems get more and more clumpy with time. With gravity, inhomogeneity is the new homogeneity. Since low-to-high entropy transitions express the natural tendency of systems, it ought to be that in gravitating systems clumped states are of high entropy and spread out ones of low entropy. The cosmic background radiation shows that the universe was more homogeneous in the past. Hence the past state is vindicated. In fact, one might go so far as to say not only that it does not falsify the standard explanation of entropy increase, but that it is a stunningly accurate prediction made by the standard explanation.

Of course, this simple idea leaves out the momentum sector of phase space. There is no 'natural tendency' toward *spatial* homogeneity or inhomogeneity in either gravitating or non-gravitating systems. The oil and vinegar separating in your container of salad dressing is an entropy increasing process. Many spatial inhomogeneities grow in perfectly normal entropy increasing situations, and presumably homogeneities can develop in gravitational situations. The idea must be, then, that the increasing concentration in the configuration sector of phase space is compensated by a greater decreasing concentration in the momentum sector of phase space. As we go forward in time, one – not implausibly – imagines the velocity vectors as becoming increasingly chaotic.

Assume that the total entropy can be expressed as a simple sum of the configuration sector entropy $|\Gamma_{Mq}|$ and the momentum sector entropy $|\Gamma_{Mp}|$. Then it is easy to see that it is possible that entropy increase or decrease with time. When gravity is the dominant force then presumably $|\Gamma_{Mq}|$ will decrease as time passes if the initial state is originally very dispersed. Although the system may develop into various quasi-stable configurations, in the long run we might expect it to become more concentrated in space. On the other hand, we might expect particles gradually to be 'slingshot' far away, so that the system evaporates and becomes very dispersed in space. Similarly, it is possible that $|\Gamma_{Mp}|$ grows as particles' velocities become increasingly randomly distributed with time; but it's also possible that the velocities become more aligned as time passes. What is needed is that $\log |\Gamma_M|$ increases with time and this can be achieved in a variety of ways.

We know in the normal non-gravitational case that entropy can go up or down. There are, as described before, good and bad initial states, the bad ones leading to subsequent entropy decrease. Fortunately, with respect to the Lebesgue measure, most of the states are believed to be good ones. So what is of interest is not whether entropy can go up or down when gravity is turned on – of course it can – but whether for *most* initial states entropy increases.

A really pressing question then is whether the standard probability distribution crafted from the Lebesgue measure is *empirically adequate* when gravitational interactions are included. Can we see the motion of the stars, and so on, as the movement to an equilibrium state, where equilibrium is understood as the largest, 'most probable' macro-state according to the Lebesgue measure? A priori, the applicability of the Boltzmann framework, and in particular, the empirical adequacy of the Maxwell–Boltzmann probability distribution, is not guaranteed. New physics presents new challenges. Indeed, when one states this hypothesis one realizes that the standard explanation of the direction of time – which assumes this framework works with gravity – tries to explain with one stroke two possibly quite distinct processes. It tries to account for ordinary thermodynamics *and* the rise of structure in the cosmos. The first is a primarily non-gravitational process and the second is a primarily gravitational process. The past hypothesis is thus a tremendously ambitious claim, and if successful, the result would be a major unification in physics. But we should be clear that it is ambitious and that it's not obvious that the two processes can be given the same explanation.[6]

At this point the natural thing to do would be to calculate the Boltzmann entropy, with gravity included, of some toy gravitational systems and see if entropy increases. Then one would like to compare the results there with our actual cosmological history. However, for various reasons to be discussed, we are stymied in this attempt.

## 3.4 The gravitational paradoxes

Statistical mechanics and thermodynamics work flawlessly in some gravitational contexts. In terrestrial cases where we can approximate the gravitational field as

---

[6] I should not give the impression that no one else is aware of potential difficulties with the usual response besides Earman. Wald (2006), for instance, comments that statistical thermodynamics is usually justified via ergodicity, and yet ergodicity will not obtain in a general relativistic universe (the universe might be open, and it is not time translation invariant in the right way). He also warns that the real story will include discussion of black hole entropy and quantum gravity. As mentioned above, I think that unless one shows that the black hole entropy is connected to the Boltzmann entropy, then the black hole entropy will not be relevant to our explanation. The first worry may also be irrelevant, as stated, since the Boltzmannian hopes his or her explanation uses requirements on the dynamics that are weaker than ergodicity. But the spirit of Wald's point is right: once the Boltzmannian is clear about the necessary dynamics, it will be a good question whether they obtain in generally relativistic spacetimes.

uniform, there is simply no problem. Thermodynamics obviously works in such cases, and the extension of statistical mechanics to systems with external uniform fields does not require any major modification (see, e.g., Landau and Lifshitz, 1969: 72; Rowlinson, 1993). However, we are interested in the more general question, the thermodynamic and statistical mechanical properties of a system self-interacting via time-varying gravitational forces. We will operate under the idealization that gravity is the only force obtaining between the particles and we will restrict ourselves to classical gravitation theory. Therefore, we investigate the thermodynamic properties of the famous classical $N$-body problem in gravitation theory.

The Hamiltonian describing the system for $N$ gravitationally interacting particles of mass **m** is

$$H(q, p) = \sum_{i=1}^{N} \frac{p_i^2}{2m_i} - \frac{1}{2} \sum_{i=1}^{N} \sum_{j \neq i} \frac{Gm_i m_j}{r_{ij}} \qquad (3.1)$$

where $r_{ij} = \|q_i - q_j\|$. Although this system is ideal, some globular star clusters ($N = 10^5$–$10^6$), galaxies ($N = 10^6$–$10^{12}$), open clusters ($N = 10^2$–$10^4$), and planetary systems are decent instantiations of this ideal system. That is, the salient features of these systems over large time and space scales are due to gravity. Collisions, close encounters and other behaviours where other forces are relevant are rare, so ignoring inelastic interactions does not cause great harm. The question is then whether the stars in such systems or even the galaxies themselves, when idealized as point particles, admit a thermodynamic description. Can we think of the stars as the point particles in a thermodynamic gas?

In the literature on classical gravitational thermodynamics,[7] most papers mention some subset of five obstacles facing any such theory: non-extensivity, ultraviolet divergence, infrared divergence, lack of equilibrium, and negative heat capacity. The first problem is that the energy and entropy of systems evolving according to (3.1) can be non-extensive, even though in thermodynamics these quantities are extensive. The second problem arises from the infinite range of the gravitational potential and the lack of gravitational shielding; together they imply that the integral over the density of states can diverge. The third problem arises instead from the short-range nature of the potential. Here the problem is the local singularity of the Newtonian pair interaction potential. Two classical point particles can move arbitrarily close to one another. As they do so, they release infinite negative gravitational potential energy. Partition functions, which need to sum over all these states, thereby diverge. The fourth problem comes in many forms, some linked to

---

[7] For entries into this literature, see Padmanabhan (1990); Saslaw (2000); Dauxois *et al.* (2002); Heggie and Hut (2003).

the divergence problems. But in general there are many problems with defining an equilibrium state for a system evolving via (3.1). Finally, the fifth problem, which is not really a paradox but merely extremely counterintuitive, is that the heat capacity for systems evolving via (3.1) can often be negative, whereas in classical thermodynamics it is always positive.

To get an intuitive feel for how gravity causes trouble, focus on just one issue, the non-extensivity problem. Intuitively put, extensive quantities are those that depend upon the amount of material or size of the system, whereas intensive quantities are those that do not. The mass, internal energy, entropy, volume and various thermodynamic potentials (e.g., $F, G, H$) are examples of extensive variables. The density, temperature, and pressure are examples of intensive variables. Mathematically, the most common expression for extensivity is the definition that a function $f$ of thermodynamic variables is extensive if it is homogeneous of degree one. If we consider a function of the internal energy $U$, volume $V$, and particle number $N$, homogeneity of degree one means that

$$f(aU, aV, aN) = af(U, V, N) \qquad (3.2)$$

for all positive numbers $a$. Consider a box of gas in equilibrium with a partition in the middle and consider the entropy, so that $a = 2$ and $f = S$. Then $S(2U, 2V, 2N)$ represents the joint system, and Equation (3.2) says that this is the same as two times the individual entropies of the partitioned component systems. Extensive functions are also assumed to be additive, and with a slight assumption, they are. A function – using our example, entropy – is additive if

$$S(U_1 + U_2, V_1 + V_2, N_1 + N_2) = S(U_1 + V_1 + N_1) + S(U_2 + V_2 + N_2)$$

With minimal assumptions homogeneity of degree one implies additivity.[8] With these definitions in hand, let us turn to statistical mechanics, the theory that explains why thermodynamical relationships hold for mechanical systems.

Perhaps the most basic assumption of thermodynamics / statistical mechanics is that the total energy of any thermodynamic system is approximately equal to the sum of the energies of that system's subsystems. If we have a large gas in a box, and we conceptually divide it into two subsystems, we expect the total energy to be the sum of the two subsystem energies – so long as the subsystems are still macroscopic systems. In many influential treatments of the theory, this assumption is regarded as the most basic of all, e.g., Landau and Lifshitz's (1969) classic treatment begins with essentially this assumption.

One of the features that makes this assumption plausible is that in terrestrial cases we are usually dealing with short-range potentials. At a certain scale matter

---

[8] See Dunning-Davies (1983) and Touchette (2002) for useful discussions of extensivity and additivity. Because of their close connections for realistic systems, I will use the two more or less interchangeably.

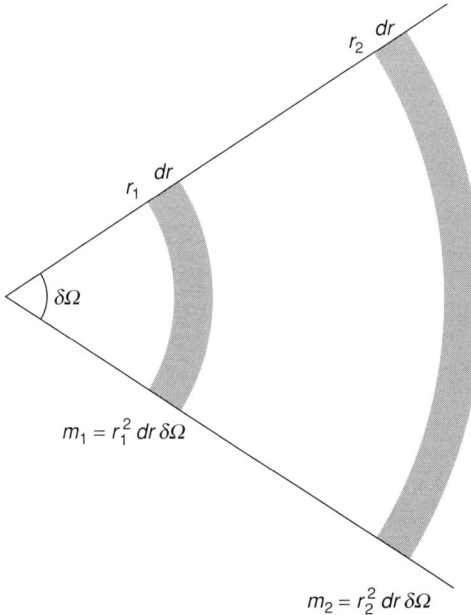

Fig. 3.1. Fashioned after Binney and Tremaine (1987: 188)

is electrically neutral and gravity is so weak as to be insignificant. If the potential is short range and our subsystems are not too small, then the subsystems will interact with one another only at or in the neighbourhood of their boundaries. When we add up the energies of the subsystems, we ignore these interaction energies. The justification for this is that the interaction energies are proportional to the *surfaces* of the subsystems, whereas the subsystem energies are proportional to the *volumes* of the subsystems. So long as the subsystems are big enough, the subsystem energies will vastly trump the interaction energies as the number of subsystems increases because the former scale as (length)$^3$ and the latter as (length)$^2$. The basic assumption is then justified.

However, if gas molecules are replaced by stars – that is, short-range potentials replaced by long-range potentials – this reasoning does not work. Consider a star at the apex of a cone (Binney and Tremain, 1987: 187–8) and the force by which the stars in the cone attract the star (Fig. 3.1). Suppose the other stars are distributed with a uniform density. The force between this star and any other falls off as $r^{-2}$, but the number of such stars increases along the length of the cone as $r^2$. Thus any two equal lengths of the cone will attract the target star with equal force. If the density of stars is perfectly homogeneous and isotropic, the star will not feel any force. But if not homogeneous – even if not homogeneous only at great distances – the star will feel a net force. For this reason the force on any particular star is typically determined more by the gross distribution of matter in the galaxy than by the stars close to it. Collisions do not play as large a role as they do in a typical gas in a box on Earth. As is sometimes said, terrestrial gas molecules tend to lead violent lives determined in large part by sudden disputes with their neighbours; stars tend to lead comparatively peaceful lives because they are in harmony with the overall universe.

Returning to energy, we see that the interaction energies may not be proportional to the subsystem surfaces. For short-range potentials, the dominant contribution to the energy comes from nearby particles; but for long-range potentials, the dominant contribution comes from distant particles. To drive home the point, consider a sphere filled with a uniform distribution of particles. Now add a particle to the origin and consider its internal energy U:

$$U \propto \int_0^R 4\pi r^2 \, dr \, \rho r^{-3-\varepsilon} \propto \int_0^R dr \, r^{-1-\varepsilon}$$

One can then verify that with $\varepsilon > 0$ ('short-range potentials'), the significant contribution to the integral comes from near the particle's origin, whereas with $\varepsilon < 0$ ('long-range potentials'), the contribution comes from far from the origin (Padmanabhan, 1990).

Consider again a chamber of gas divided into two equal boxes, A and B. If the particles are interacting via long-range forces, the particles in box A will feel the particles in box B as much or even more than the particles nearby. Let $E_A$ represent the energy of box A and $E_B$ represent the energy of box B. As a result of the interaction, it is easy to devise scenarios whereby $E_A = E_B = -a$, where $a > 0$, yet where the energy of the combined system $E = 0$, not $-2a$. The energy might not be even approximately additive.

When the additivity and extensivity go, so do large portions of equilibrium thermodynamics and statistical mechanics. For example, when a system is in equilibrium, its large subsystems also will be in equilibrium. This is no longer necessarily the case. And additivity is a requirement for the equilibrium second law in thermodynamics (see Lieb and Yngvason, 1998). Moreover, in statistical mechanics it is built into the heart of the theory. The famous Boltzmannian probability $W$ of a macro-state is assumed equal to the product of the probabilities of the subsystem macro-states, i.e. $W_{total} = W_a W_b$. Boltzmann's definition of entropy as $S = k \ln W$ straightforwardly implies that $S_{total} = S_a + S_b$. And there is no conventional thermodynamic limit for non-extensive systems. For a rigorous discussion of this see Padmanabhan (1990), Lévy-Leblond (1969) and Hertel *et al.* (1972). This last fact should not be surprising. The existence of the thermodynamic limit depends on making the contribution of surface effects go to zero as $N$, $V$ go to infinity. In a non-additive system, we saw that the surface effects are not going to get smaller as *N* and *V* increase. Ironically, the thermodynamic limit does not apply to very large systems if one includes the force primarily relevant to the dynamics of those systems.[9]

## 3.5 The problem

As interesting as these problems are, they are – at first glance – orthogonal to our main worry. The problems described are problems for *equilibrium* thermodynamics and statistical mechanics, but we are interested in *non-equilibrium* statistical mechanics.

The reason this is so is because the past state is surely not an equilibrium state, yet arguably it still has a Boltzmann entropy. Why is the past state a non-equilibrium state? It is the global state of the universe, and the very reason it is posited is that subsequent evolution will spontaneously take the global state to regions corresponding to higher entropy. But if it is in equilibrium, the system will not change unless an external constraint is removed; yet since the system is the global state, there is no external constraint to remove. The unavoidable conclusion is that the past state must not be an equilibrium state. Indeed, no one expects the past state to stay that way. It is expected, under the attractive force of gravity, to begin clumping. The past state, therefore, simply does not have an ordinary equilibrium entropy corresponding to equilibrium thermodynamics or equilibrium statistical mechanics. But it does have a Boltzmann entropy. The definability of the Boltzmann entropy in systems outside equilibrium is touted as perhaps its greatest virtue. Since we are restricting ourselves to the classical phase space and assuming a Lebesgue measure upon it, we do not have the potentially in-principle problems Earman worries about with general relativity. All one really needs for a Boltzmann entropy to exist is a well-defined macro-state and a well-defined notion of volume in phase space. If the earliest cosmological times do not correspond to a macro-state for some reason, then the past hypothesis picks out the 'first' state that does. This macro-state will correspond to a particular volume $|\Gamma_M|$, and hence it has an entropy. The problems with equilibrium statistical mechanics in the presence of

---

[9] Before concluding this section I should point out that there is a large research program devoted to the statistics of non-extensive systems that I am here bracketing aside. This is the approach of Tsallis statistics. The Tsallis school develops a generalization of the Boltzmann and Gibbs entropies, namely, the Tsallis entropy. The Tsallis entropy reduces to the Boltzmann and Gibbs entropies when the system is extensive, but is different otherwise. The motivation behind the program is to show that the Tsallis entropy works well in situations where the Boltzmann and Gibbs entropies allegedly break down. Long-range force systems like self-gravitating systems are supposed to be one example. The debate between the Tsallis school and others believing Boltzmann–Gibbs suffices is often very heated. For the purposes of this chapter I want to stay conservative and remain within the Boltzmann framework – though for some criticism of the Tsallis school, see Nauenberg (2003). That said, we ought to acknowledge that one way of responding to the above worries is to change frameworks and go outside the normal Boltzmann–Gibbs picture.

gravity are worrying, but so far not directly relevant to the increase of Boltzmann entropy.

Or are they? One cannot completely divorce non-equilibrium theory from equilibrium theory. Think of the issue as follows. The Boltzmann entropy for the gravitating system described by the past state will exist, but what will it do? What one wants is not merely the existence of entropy but also the functional relationships that are usually entailed by a system having an entropy. Why think, for instance, that a system in the past state will increase its entropy? And more generally, granted that the past state picks out some volume in phase space, what gives this volume its physical significance?

As Boltzmann famously showed, in the case of the dilute gas we have everything we could want from the Boltzmann entropy. Recall that the argument goes as follows. The $H$-theorem shows that the entropy $S(f(x, v))$

$$S(f) = -N \int f \ln f \, d^3\mathbf{x} \, d^3\mathbf{v} \tag{3.3}$$

increases monotonically with time when $f$ is evolving via the (independently motivated) Boltzmann equation. Here $f$ is the distribution function defined over six-dimensional $\mu$-space when partitioned into a finite number of cells. $S(f)$ is shown to increase with time except for when $f$ is a local Maxwellian, whereupon $S(f)$ is stationary. Since Maxwell had already shown that his distribution corresponded to equilibrium, the idea of $S(f)$ playing the role of entropy is naturally suggested. Note that so far none of this bears on the Boltzmann entropy. The crucial link is provided by the detour via six-dimensional $\mu$-space. By making a number of assumptions appropriate to the dilute gas – but certainly not to gases with strong interactions – Boltzmann is able to 'translate' distributions $f$ into hypervolumes in $\Gamma$. In particular, he is able to show via the famous 'combinatorial argument' that the distribution $f$ corresponding to the Maxwell distribution occupies far and away the greatest proportion of volume in $\Gamma$. Via this translation Boltzmann shows that all the desirable properties true of $S(f)$ are true of the Boltzmann entropy too in the case of the dilute gas. Doing so motivates the entire picture of micro-states most likely evolving into the dominant equilibrium sections of $\Gamma$. (See Uffink, 2007 for more discussion.)

It is important to stress that it is the above connections to the $H$-function and the Boltzmann equation that gives the volume in $\Gamma$ any claim to be physically significant. After all, there are other volumes calculated in other bases, e.g., energy, which do not have this feature.

Now we immediately see at least one big problem for providing the Boltzmann entropy physical significance in the gravitational case. Boltzmann's argument can plausibly be extended to some systems for which it was not originally intended, and new arguments mimicking Boltzmann's can show that the Boltzmann entropy for some non-dilute gases has physical significance (e.g, Garrido *et al.*, 2004; Goldstein and Lebowitz, 2004). However, in the gravitational case we know we in general cannot use Boltzmann's argument and there is not much reason to hope anything like it will help.

For instance, consider an important property we need to know of our system: the macro-state $f(\mathbf{x}, \mathbf{v})$ that has maximum volume in $\Gamma$. One can hope to find this via the combinatorial argument only if one can translate between $\mu$-space and $\Gamma$-space – and one can only do this because the gas is dilute and interactions are effectively turned off. What one does is maximize $f(\mathbf{x}, \mathbf{v})$ subject to two constraints. One constraint is associated with particle number, but the other is more directly relevant to us:

$$\int_V d\mathbf{x} \int_{R^3} d\mathbf{v}\, \frac{1}{2} m\mathbf{v}^2 f(\mathbf{x}, \mathbf{v}) = E \qquad (3.4)$$

where $E$ is the total energy. In other words, one is maximizing conditional on the claim that the total energy is the sum of kinetic energies. If this is so, the Maxwell equilibrium distribution is the macro-state with the maximum volume in $\Gamma$. In fact, as $n$ goes to infinity departures from the equilibrium macro-state go to zero. This step warrants the additional geometric interpretation the Boltzmannian asserts. The picture of typical non-equilibrium states moving to equilibrium states because there are vastly more of the latter than former is not justified except by this procedure.

In the present case, however, we are directly challenged by the total energy not being approximately the sum of independent individual energies. Equation (3.4) is manifestly false and not even approximately true for many self-gravitating systems. The gravitational potential energy contributes to the overall energy of the system. Without (3.4) one cannot show that the largest macro-state in phase space is the equilibrium state; absent this, one cannot make plausible that typical initial states go to equilibrium. So although the loss of a maximization constraint may seem like a quibble, an awful lot hangs on it. In fact, the very terms by which we conceived the original question depend on this; unless the energy factorizes there is no reason to think the entropy $S(f)$ is a simple sum of a configurational contribution and momenta contribution, so the intuitive reasoning we engaged in earlier does not hold. And if this were not bad enough, we are also lacking a gravitational version of the Boltzmann equation for which one can prove an $H$-theorem (more on this later). I hope this discussion adequately displays the problem: although the gravitational system has a well-defined Boltzmann entropy, that by itself does not imply any particular subsequent behaviour.

Perhaps we can look at the glass as half full? We already knew the above problems for the Boltzmann explanation. Many critics of Boltzmann (e.g. Schrödinger, 1948 [1989]) point out that it works rigorously only for the case of dilute gases, yet most systems are of course not dilute gases. The Boltzmannian can deflate some of these worries by showing how many systems are approximately like dilute gases, how numerical simulations of cases that are not dilute gases vindicate the Boltzmannian claims, and so on. But there are of course many systems that do not fit this mould, and the strongly self-gravitational system is one of them. All we have done is highlight the existing problem by displaying a class of systems that are especially far from being treated as dilute gases. And we could have made this argument with plenty of non-gravitational systems too, e.g., some types of plasmas. Maybe, perversely, this is good news to the Boltzmannian. The problem gravitational interaction presents to the standard story the Boltzmannian tells is *as bad as* but *not obviously worse than* the problem other systems already cause the Boltzmannian.

It would be nice if we could view the problem as simply a new version of the same old one already challenging Boltzmann. But it's not clear that even this is the case. In astrophysics researchers often make assumptions about the stars that warrant a description of the system via $f$ on $\mu$-space, not the full $\Gamma$-space. That is, they often work with the 'one-particle' distribution function on six-dimensional $\mu$-space just as Boltzmann did in his work on dilute gases. This restriction on $f$ is typically justified in the astrophysics literature by the fact that gravitational systems are essentially collisionless for long periods of time. So what we are doing now is restricting ourselves to a regime wherein some of the usual Boltzmann apparatus can be salvaged. The entropy is defined as (3.3) above.

Under these restrictions, let us now search for the state of maximum entropy, which will be our equilibrium state. Even here, however, we run into problems. To find out what equilibrium looks like for self-gravitating systems, therefore, we can find the distribution $f(\mathbf{x}, \mathbf{v})$ that maximizes the equilibrium entropy (3.3). However, if one looks for the distribution that maximizes $S$ for a given mass $M$ and energy $E$, then it is a major result in the field that $S$ is extremized iff $f(\mathbf{x}, \mathbf{v})$ is the distribution function of the isothermal sphere (Ogorodnikov, 1965; Lynden-Bell, 1967; Lynden-Bell and Wood, 1968). The isothermal sphere is an infinite $n$ self-gravitating ideal gas. That is, there is *no* distribution function that maximizes $S$ while keeping $M$ and $E$ finite. Maintaining finite $M$ and $E$, one can obtain arbitrarily large entropies by rearranging the configuration of stars, as Binney and Tremaine (1987) show. There is no $f(\mathbf{x}, \mathbf{v})$ that maximizes the entropy (3.3) for finite $M$ and $E$. (Binney and Tremaine (1987) take this result to show that galaxies and presumably other typical stellar configurations are *not* the result of long-term thermal equilibrium.

The quest in the astrophysics literature is to associate typical stellar configurations with quasi-stationary states, not true equilibrium states.[10])

The Boltzmannian may reply that this problem is an artefact of the simplification, that with the 'true physics' on $\Gamma$ (instead of on $\mu$) the problem will go away. That puts the Boltzmannian in an awkward position, however. The Boltzmannian cannot show this is the case because then she meets the gravitational version of Schrödinger's worry: that one cannot prove much outside the simple case. In the non-gravitational case the Boltzmannian replies to Schrödinger by pointing out all the success she had with dilute gases, toy models, computer simulations, and so on. Now in the gravitational case it looks like the Boltzmannian needs to solve the hard case to help answer problems with the allegedly easy case.

Obviously more study is needed of this problem. Perhaps there is still a way the Boltzmannian can by-pass these difficulties. Right now, however, gravity seems to have pulled the Boltzmannian into a serious thicket of problems.

## 3.6 A way forward?

That was the bad news. Let us conclude, however, with some good news.

In Section 3.3 we learned that the problem we have is giving the Boltzmann entropy of a gravitating system physical meaning. In the case of a non-gravitating dilute gas we saw that the Boltzmann equation, the $H$-theorem and the connection to the $H$-function provided the Boltzmann entropy physical significance. Can we do this for other systems, in particular, gravitational systems?

In some regimes, yes.

To understand the general idea, recall again what Boltzmann does for dilute gases. Boltzmann considers a distribution $f(\mathbf{x}, \mathbf{v}, t)$ that evolves according to the Boltzmann equation, an equation independently motivated on physical grounds. His famous $H$-theorem shows that a function of $f$, $S(f)$, increases with time except when $f(\mathbf{x}, \mathbf{v})$ is a local Maxwellian. The reason why any of this is interesting is that $S(f)$ is shown to be roughly equal to the Bolzmann entropy $k \log |\Gamma_M(X)|$ and the Maxwellian distribution is the distribution corresponding to the maximum Boltzmann entropy. Thanks to this connection, we know that so long as the system is appropriately modelled via the Boltzmann equation, the genuine Boltzmann entropy will increase with time until reaching equilibrium.

There are physical regimes, of course, where the Boltzmann equation is not a good approximation, and in those cases different macroscopic kinetic equations often apply. Are there macroscopic kinetic equations that are good approximations

---

[10] See, for instance, Chavanis (2005).

of real systems wherein the dominant interaction is gravitational? Can we get an *H*-theorem for these regimes? And can we show that this *H*-function $H = -S(f)$ is roughly equal to the Boltzmann entropy? The answer in each case is yes.

In astrophysics the full *n*-body problem is often too hard to study, even in computer simulations in many instances, and hence one considers regimes wherein various kinetic equations are appropriate. Astrophysics is filled with macroscopic equations of motion for distributions. Indeed, since plasmas interact via long-range forces too, many of the same kinetic equations used in plasma physics often work in astrophysics too, so an awful lot is known about many equations.

Consider a galaxy of *n* identical stars with characteristic radius *r*. The time it takes any star to cross the galaxy is *r*/*v*, where *v* is the typical speed of a star (determined by *G*, *n*, *r*, mass); this is called the *crossing time* of the star. Suppose that the star evolves in a background wherein the mass is perfectly smoothly distributed, not clumped up into individual stars. Call this its mean trajectory. When would the difference between the star evolving in this background versus a more realistic background show up in its velocity (where by 'show up' we mean the velocity changes by order of itself)? Leaving the details to textbooks on galactic dynamics, the answer is that the star is deflected from its mean trajectory over order 0.1 *n*/ln *n* crossing times. Hence one concludes that for systems that are less than 0.1 *n*/ln *n* crossing times old, individual stellar encounters are more or less unimportant. Many galaxies, with $n \approx 10^{11}$ stars and a few hundred crossing times old, are examples. For these systems, typically where the forces are long range and weak, a natural move is to replace the actual force by its spatial average. Many self-gravitating systems enjoy large space and time scales where this approximation is justified.

A major equation of study in galactic dynamics is therefore the Vlasov equation, or the collisionless Boltzmann equation. The equation is for a density of particles subject to an average force field:

$$\frac{\partial f}{\partial t} + \mathbf{v} \cdot \nabla f - \nabla \Phi \cdot \frac{\partial f}{\partial \mathbf{v}} = 0 \qquad (3.5)$$

where $f = f(\mathbf{x}, \mathbf{v}, t) d^3\mathbf{x}\, d^3\mathbf{v}$ and $\phi$ is a smooth gravitational potential. Equation (3.5) is essentially a special case of the Liouville equation (for a derivation of (3.5), see Kandrup, ms; 1981). Despite its simplicity, the Vlasov equation is described in textbooks as the fundamental equation of stellar dynamics. What is nice for us is that if we define an entropy via this *f*, $S(f)$, then one can show that it is proportional to the Boltzmann entropy. What is not so nice, however, is that we cannot show that entropy increases for distributions evolving according to the Vlasov equation.[11]

---

[11] In terms of the conjecture mentioned in Section 3.2, DeRoeck *et al.* (2006) make clear that not all macroscopic equations will produce an *H*-theorem. In particular, and skipping the details, they explain that if *every* microstate *X* is typical of the macroscopic equation, then the argument does not go through. For Equation (3.5),

This is not at all surprising, since the Vlasov equation is more or less the Liouville equation.[12]

Nonetheless, there is a lot more to stellar dynamics than the Vlasov equation. Many systems are such that stellar encounters have played a major part in their development. Globular clusters, open clusters, galactic nuclei and clusters of galaxies all have $n$, crossing times and lifetimes making the collisionless regime inappropriate to describe them. Outside this Vlasov regime, kinetic equations other than (3.5) are required, equations including some effect of collisions and close encounters. There are scores of kinetic equations used in the subject, but for concreteness let me mention two, namely, the Fokker–Planck equation and the essentially equivalent Landau equation. (The Landau equation is a symmetric form of the Fokker–Planck equation.) These are equations of form

$$\frac{\partial f}{\partial t} + \mathbf{v} \cdot \nabla f - \nabla \Phi \cdot \frac{\partial f}{\partial \mathbf{v}} = C[f] \qquad (3.6)$$

where $C[f]$ denotes the rate of change of $f$ due to encounters and collisions. The Fokker–Planck and Landau equations are of form (3.6) with specific collision terms derived for small-angle grazing collisions. The equations are derived by expanding the Boltzmann equation about small-angle grazing collisions. For the exact form of $C[f]$ see Balescu (1963), Spohn (1991) or Heggie and Hut (2003). Both Fokker–Planck and Landau are useful for gas/fluid systems that are weakly coupled, and they are particular useful for stellar systems in which collisions are rare and interactions weak. In astrophysics, the Fokker–Planck equation is advertised as the 'most accurate model of a stellar system, short of the N-body model' (Heggie and Hut, 2003: 87).

What I want to point out is that for the Fokker–Planck equation, one of the most successful kinetic equations in astrophysics, one can get everything one wants. In particular, for many broad classes of collision terms $C[f]$ one can prove an $H$-theorem for (3.6). One can show that this $H$-function is related to the Boltzmann entropy in the same way Boltzmann does for the dilute gases. And one can show that the stationary or equilibrium distribution of (3.6) is equivalent to the solution one obtains from maximizing the Boltzmann entropy in the presence of an external potential. Since the Fokker–Planck equation has been extensively studied, and these results are relatively well known, I will not prove any of it here. I simply will refer the reader to the relevant literature for proofs and discussions of these assertions (see, e.g., Green, 1952; Balescu, 1963: 170ff; Liboff and Fedele, 1967; Spohn, 1991: 83;

---

every $X$ is typical: every solution of Hamilton's equations will follow solutions of (3.5) for $f$. We will not, therefore, get an $H$-theorem.

[12] A coarse-grained entropy might increase, however. In gravitational dynamics physicists speak of non-collisional 'phase mixing' as another means of a system moving to equilibrium. See Heggie and Hut (2003: 93) and Chavanis (1998).

van Kampen, 1981; Risken, 1989).[13] I note in addition that many of these results have recently been extended to the non-linear Fokker–Planck equation too (e.g., Frank (2005) and references therein). Of course, complete vindication of my claim will hang on demonstrating the match between particular astrophysical systems and the assumptions (boundary conditions and so on) used in any particular $H$-thoerem.

There are scores of other kinetic equations used in astrophysics and for many of these one will also find an $H$-theorem in the literature. And for those that do not readily admit of an $H$-theorem, one may also try employing the conjecture of De Roeck *et al.* to find an '$H$-theorem' of sorts. Recall that in Section 3.2 I described a top–down way of thinking about entropy increase and $H$-theorems. Imagine we have some deterministic macroscopic equation of motion, one that tells us that macro-states such as $M_1$ at $t_1$ will evolve by time $t_2$ into macro-state $M_2$. We saw that Liouville's theorem and the claim that (effectively) $\phi_{t_2-t_1} \Gamma_{M_1 t_1} \subset \Gamma_{M_2 t_2}$ implies that almost all of the points originally in $M_1$ have evolved into the set corresponding to $M_2$. From this it follows that $\Gamma_{M_1 t_1} \leq \Gamma_{M_2 t_2}$ and we therefore have a kind of $H$-theorem. Whether this strategy is defensible and whether it works with certain equations in astrophysics are questions that require study. I will not argue for either here. Presently I merely wish to point out that with the plethora of macroscopic kinetic equations in the field, there will be many opportunities to try to employ this strategy.

The picture we have developed, then, is this. We have not calculated the Boltzmann entropy including strong gravitational coupling directly, so we do not know whether it increases or decreases from an initial state like the early cosmological state. For the reasons discussed in Section 3.3, unless we simplify our system considerably we cannot show what the Boltzmann entropy of such a state will do. We have no answer to the main question of this chapter; indeed, displaying this problem is the main point of the chapter. As mentioned, however, the news is not all bad. We know that when some large self-gravitating structures in the universe reach a certain stage of development it becomes appropriate to idealize them as obeying a gravitational kinetic equation. For some of these equations, and in fact for some very accurate ones, we can show that the Boltzmann entropy increases. I have not shown this here, as it is implicit in the literature.[14] Moreover, I have

---

[13] Please bear in mind that often these works are not written from the perspective of the Boltzmann viewpoint used here. To complete all the links mentioned, one sometimes will need to use, for example, the fact that the Boltzmann entropy is close in value to the Gibbs entropy at equilibrium for large systems, as well as results from Boltzmann's original derivation.

[14] It may be worth pointing out that the diffusion coefficient in the Fokker–Planck equation causes dispersal in velocity space. So if we think back to Section 3.3, where we wanted to know what was happening in momentum space in such systems, we see that these kinetic equations are describing systems whose momenta are getting more dispersed as time goes on, just as we hoped.

pointed to the vast range of gravitational kinetic equations in use as a place to investigate this question further.

To what extent the Boltzmannian program is ultimately successful in the face of gravity depends on what we hope for and on the empirical facts. The original past hypothesis covered the entire universe, but this theory will not be vindicated by the current very limited result. The current move only yields an increasing Boltzmann entropy in regimes appropriately described by a gravitational kinetic equation. For instance, the Fokker–Planck regime only lasts when the system is weakly coupled. The whole universe is certainly not such a regime. If one hopes for a Boltzmann entropy for the universe, this avenue cannot meet this goal. Also, if one wanted to tackle the problems of extensivity *et al.* head-on, we have not done that here either. By going to a regime where a mean force is used, even where close encounters are considered to some extent, we may be accused of ignoring the problem of extensivity rather than addressing it.

Yet if one has more modest expectations, one has encouraging news. What is perhaps the best kinetic equation incorporating gravitational effects generates the increase of Boltzmann entropy. The natural reconstruction of the past hypothesis is as the claim that the early states of (e.g.) Fokker–Planck regimes are of very low Boltzmann entropy compared to now. The pressing empirical question for this approach is whether we are in such a regime and if so how big it is and how many there are.[15]

This picture, it must be said, bears some similarity to the 'branch' systems approach to statistical thermodynamics. Reichenbach's (1999) branch hypothesis is the claim that thermodynamics applies only to quasi-isolated macroscopic 'branch' systems. Thermodynamics does not apply to the universe as a whole on this view, but only to certain systems when they become sufficiently isolated from the rest of the world. Historically, one objection to this picture is that it is not at all clear what 'sufficiently isolated' could possibly mean. See Albert (2000: 88–89) for a forceful statement of this objection, among others. Here I just want to note that the proposal under review is not guilty of *this* mistake, at least on one reading. The criterion of whether a system fits the assumptions underlying the use of a Fokker–Planck equation is quite clear. The identification of branches can proceed without too much difficulty. The larger problem, also mentioned by Albert, of whether one has any

---

[15] Actually, probably more of the action will come in looking at the level of detail – i.e. the choice of macro-states – than simply the size of the system. For instance, our galaxy, the Milky Way, has approximately $N = 10^{10}$ stars in it and a 'crossing time' of $10^8$ yr, making stellar close encounters a relatively unimportant part of its evolution. This means the Vlasov equation is a good description of our galaxy. This equation, recall, provides no entropy increase. However, that does not mean that if one wants to look at more fine-grained structure in our galaxy one cannot use the Fokker–Planck equation, an equation from which one can derive entropy increase. And that does not mean that one cannot also enlarge the scale and use the Fokker–Planck equation to describe the dynamics of clusters of galaxies, with $N = 10^3$, which may include the Milky Way.

right to impose a uniform probability distribution over the 'first' such state when we know it has evolved from previous states lingers, however, and demands further thought (for a little in this direction, see Callender, 2010).

In sum, I hope to have shown how the inclusion of gravity into the Boltzmannian account of the direction of time is highly ambitious but also non-trivial. After sketching the serious problems with gravity, I made plausible a sketch of how one can obtain an increasing Boltzmann entropy in self-gravitating systems described by certain types of gravitational kinetic equations. Further work is needed to judge whether this kind of approach is best, but I do hope it removes some of the pessimism one might (reasonably) have about Boltzmann's non-equilibrium framework in the presence of gravitation.[16]

## References

Albert, D. Z. (2000). *Time and Chance*. Cambridge, MA: Harvard University Press.
Balescu, R. (1963). *Statistical Mechanics of Charged Particles*. New York: Interscience Publishers.
Binney, J. J. and Tremaine, S. (1987). *Galactic Dynamics*. Princeton, NJ: Princeton University Press.
Callender, C. (2004). Measures, explanations and the past: should 'special' initial conditions be explained? *The British Journal for the Philosophy of Science*, **55**(2), 195–217.
Callender, C. (2010). The past histories of molecules. In *Probabilities in Physics*, ed. C. Beisbart and S. Hartmann. Oxford: Oxford University Press.
Chavanis, P. H. (1998). On the 'coarse-grained' evolution of collisionless stellar systems. *Monthly Notices of the Royal Astronomical Society*, **300**, 981–991.
Chavanis, P. H. (2005). On the lifetime of metastable states in self-gravitating systems. *Astronomy and Astrophysics*, **432**, 117–138.
Dauxois, T., Ruffo, S., Arimondo, E. and Wilkens, M. (eds.) (2002). *Dynamics and Thermodynamics of Systems with Long Range Interactions*. Berlin: Springer-Verlag.
DeRoeck, W., Maes, C. and Netočný, K. (2006). H-theorems from macroscopic autonomous equations. *Journal of Statistical Physics*, **123**(3), 571–584.
Dunning-Davies, J. (1983). On the Meaning of Extensivity. *Physics Letters,* **94A**, 346–348.
Earman, J. (2006). The past hypothesis: not even false. *Studies in History and Philosophy of Modern Physics*, **37**(3), 399–430.
Frank, T. (2005). *Nonlinear Fokker–Planck Equations*. Berlin: Springer-Verlag.
Frigg, R. (2009). Typicality and the approach to equilibrium in Boltzmannian Statistical Mechanics. *Philosophy of Science (Suppl.)*.
Garrido, P. L., Goldstein, S. and Lebowitz, J. L. (2004). Boltzmann entropy for dense fluids not in local equilibrium. *Physical Review Letters*, **92**, 50602-1–50602-4.

---

[16] Thanks to Jonathan Cohen, Roman Frigg, Carl Hoefer, Tarun Menon, Ioan Muntean and Allan Walstad for comments and help, as well as audiences at the 2006 Philosophy of Science Association Meeting, the 2008 Reduction, Emergence and Physics Workshop in Tilburg and the 2008 Annual Meeting of the British Society for Philosophy of Science.

Goldstein, S. (2002). Boltzmann's approach to statistical mechanics. In *Chance in Physics, Foundations and Perspectives*, ed. J. Bricmont, D. Durr, M. C. Galavotti et al., Berlin: Springer-Verlag.

Goldstein, S. and Lebowitz, J. L. (2004). On the (Boltzmann) entropy of non-equilibrium systems. *Physica D: Nonlinear Phenomena*, **193**, 53–66.

Green, M. S. (1952). Markoff random processes and the statistical mechanics of time dependent phenomena. *Journal of Chemical Physics*, **20**, 1281.

Heggie, D. C. and Hut, P. (2003). *The Gravitational Million-body Problem*. Cambridge: Cambridge University Press.

Hertel, P., Narnhofer, H. and Thirring, W. (1972). Thermodynamic functions for fermions with gravostatic and electrostatic interactions. *Communications in Mathematical Physics*, **28**, 159–176.

Kandrup, H. (1981). Generalized Landau equation for a system with a self-consistent mean field–Derivation from an $N$-particle Liouville equation. *The Astrophysical Journal*, **244**, 316.

Kandrup, H. Theoretical techniques in modern galactic dynamics, unpublished class notes available at http://www.astro.ufl.edu/~galaxy/papers/.

Landau, L. D. and Lifshitz, E. (1969). *Statistical Physics*, part 1. Oxford: Pergamon.

Lavis, D. A. (2005). Boltzmann and Gibbs: an attempted reconciliation. *Studies in History and Philosophy of Modern Physics*, **36**(2), 245–273.

Lévy-Leblond, J.-M. (1969). Nonsaturation of gravitational forces. *Journal of Mathematical Physics*, **10**(5), 806–812.

Liboff, R. L. and Fedele, J. B. (1967). Properties of the Fokker–Planck equation. *Physics of Fluids*, **10**, 1391–1402.

Lieb, E. H. and Yngvason, J. (1998). A guide to entropy and the second law of thermodynamics. *Notices of the American Mathematical Society* **45**, 571–581.

Lynden-Bell, D. (1967). Statistical mechanics of violent relaxation in stellar systems. *Monthly Notices of the Royal Astronomical Society*, **136**, 101–121.

Lynden-Bell, D. and Wood, R. (1968). The gravo-thermal catastrophe in isothermal spheres and the onset of red-giant structure for stellar systems. *Monthly Notices of the Royal Astronomical Society*, **138**, 495.

Nauenberg, M. (2003). Critique of Q-entropy for thermal statistics. *Physical Review E*, **67**(3), 036114.

Ogorodnikov, K. F. (1965). *Dynamics of Stellar Systems*. New York: Pergamon.

Padmanabhan, T. (1990). Statistical mechanics of gravitating systems. *Physics Reports*, **188**(5), 285–362.

Price, H. (1996). *Time's Arrow and Archimedes' Point*. New York: Oxford University Press.

Reichenbach, H. (1999). *The Direction of Time*. Dover, NY: Mineola.

Risken, H. (1989). *The Fokker–Planck Equation: Methods of Solution and Applications*. Berlin: Springer-Verlag.

Rowlinson, J. S. (1993). Thermodynamics of inhomogeneous systems. *Pure and Applied Chemistry*, **65**, 873.

Saslaw, W. C. (2000). *The Distribution of the Galaxies: Gravitational Clustering in Cosmology*. Cambridge: Cambridge University Press.

Schrödinger, E. (1989 [1948]). *Statistical Thermodynamics*. Dover reprint.

Spohn, H. (1991). *Large Scale Dynamics of Interfacing Particles*. Berlin: Springer.

Touchette, H. (2002). When is a quantity additive, and when is it extensive? *Physica A: Statistical Mechanics and its Applications*, **305**(1–2), 84–88.

Uffink, J. (2007). Compendium of the foundations of classical statistical physics. In *Philosophy of Physics, Handbook of the Philosophy of Science*, eds. J. Butterfield and J. Earman. Amsterdam: North-Holland, pp. 923–1047.
Van Kampen (1981). *Stochastic Processes in Physics and Chemistry*. Amsterdam North-Holland.
Wald, R. M. (2006). The arrow of time and the initial conditions of the Universe. *Studies in History and Philosophy of Modern Physics,* **37**(3), 394–398.

# 4
# Quantum gravity and the arrow of time

CLAUS KIEFER

## 4.1 Arrows of time

Although most fundamental laws of physics do not distinguish between past and present, there exist classes of phenomena in Nature that exhibit a direction of time: their time-reversed versions, although in perfect accordance with the underlying laws, are under ordinary conditions never observed. Arthur Eddington called such classes 'arrows of time', see Zeh (2007) for a detailed overview. The main arrows of time are the following:

- radiation arrow (advanced versus retarded radiation);
- second law of thermodynamics (increase of entropy);
- quantum theory (measurement process and emergence of classical properties);
- gravitational phenomena (expansion of the universe and emergence of structure by gravitational condensation).

The expansion of the universe is distinguished because it does not refer to a class of phenomena; it is a single process. It has therefore been suggested that it is the common root for all other arrows of time – the 'master arrow'. As will become clear below, this does indeed seem to be the case.

The *radiation arrow* is distinguished by the fact that fields interacting with local sources are usually described by *retarded* solutions, that is, solutions which in general lead to a damping of the source. Advanced solutions are excluded; they would describe the reversed process, during which the field propagates coherently towards its source, leading to its excitation instead of damping. This holds, in fact, for all wave phenomena – a stone thrown into a pond will produce outgoing waves, but the time-reversed version of coherently ingoing water waves ejecting out a stone from the pond is never observed, although it would correspond to a valuable solution of the wave equation. In electrodynamics, a solution of Maxwell's equations can be specified by

$$A^\mu = \text{source term plus boundary term}$$
$$= A^\mu_{\text{ret}} + A^\mu_{\text{in}} = A^\mu_{\text{adv}} + A^\mu_{\text{out}}$$

where $A^\mu$ is the vector potential. The important question is then why the observed phenomena obey $A^\mu \approx A^\mu_{\text{ret}}$ or, in other words, why

$$A^\mu_{\text{in}} \approx 0 \qquad (4.1)$$

holds instead of $A^\mu_{\text{out}} \approx 0$. Equation (4.1) is called a 'Sommerfeld radiation condition'.

Today the radiation arrow of time is understood as a consequence of the second law of thermodynamics: due to the absorption properties of the material which constitutes the walls of the laboratory in which electrodynamic experiments are being performed, ingoing fields will be absorbed within a very short time and (4.1) will be fulfilled. But for the thermal properties of absorbers, the second law is responsible.

The condition (4.1) also seems to hold for the universe as a whole ('darkness of the night sky'). The so-called Olbers' paradox, according to which one should expect a bright night sky arising from infinitely many stars in an eternal universe, can be solved by noting that the universe is, in fact, not static, but has a finite age and is much too young to have enough stars for a bright night sky. This is, of course, not yet sufficient to understand the validity of (4.1) for the universe as a whole. In an early stage the universe was a hot plasma in thermal equilibrium. It is the expansion of the universe and the ensuing redshift of the radiation which are responsible for the fact that radiation has decoupled from matter and cooled to its present value of about three Kelvin – the temperature of the approximately isotropic cosmic background radiation with which the night sky 'glows'. During the expansion a strong thermal non-equilibrium could develop, which then enabled the formation of structure.

The second arrow concerns the *second law* of thermodynamics: for a closed system, entropy does not decrease. The total change of entropy is given by

$$\frac{dS}{dt} = \underbrace{\left(\frac{dS}{dt}\right)_{\text{ext}}}_{dS_{\text{ext}} = \delta Q/T} + \underbrace{\left(\frac{dS}{dt}\right)_{\text{int}}}_{\geq 0}$$

so that according to the second law the second term is non-negative. As the increase of entropy is also relevant for physiological processes, the second law is responsible for the subjective experience of irreversibility, in particular for the aging process. If applied to the universe as a whole, it would predict the increase of its total entropy, which would seem to lead to its 'heat death' (*Wärmetod*).

The laws of thermodynamics are based on microscopic statistical laws which are time-symmetric. How can the second law be derived from such laws? Certainly, purely statistical arguments are insufficient, since for each microscopic process its time-reversed version would exist with the same *a priori* probability. In fact, the highest probability would belong to a system in its thermal equilibrium state.

The second law can be derived using statistical reasoning only if a special *boundary condition of low entropy in the past* is imposed. Such a boundary condition must either be postulated from the outset or derived from a fundamental theory. Formally the increase of entropy is then described by 'master equations', see, for example, Joos *et al.* (2003). These are equations for the 'relevant' (coarse-grained) part of the system. In an open system, the entropy can, of course, decrease, provided the entropy capacity of the environment is large enough to at least compensate this decrease. This is crucial for the existence of life, and a particular efficient process in this respect is photosynthesis. The huge entropy capacity of the environment comes in this case from the high temperature gradient between the hot Sun and the cold empty space: few high-energy photons (with small entropy) arrive on Earth, while many low-energy photons (with high entropy) leave it, see, for example, Penrose (1990). Therefore, also the thermodynamic arrow of time points towards cosmology: How can gravitationally condensed objects like the Sun form in the first place? Already Ludwig Boltzmann had, towards the end of the nineteenth century, speculated that the second law has its origin in cosmology.

There is also an arrow of time in quantum mechanics. The Schrödinger equation is time-reversal invariant, but the measurement process, either through a dynamical *collapse* of the wave function, or an Everett *branching*, distinguishes a direction. Growing entanglement with other degrees of freedom leads to *decoherence*: the *irreversible* emergence of classical properties for a local system through interaction with its environment (Joos *et al.*, 2003). A particle trajectory in a bubble chamber, for example, emerges through the interaction with the atoms in the chamber. Independent of whether a real collapse of the wave function exists or not, decoherence readily explains the occurrence of an 'apparent' collapse as it is observed in experiments. Information about quantum-mechanical superpositions is delocalized into correlations with the inaccessible environment and no longer available at the system itself. Decoherence is described quantitatively through the dynamics of a reduced density matrix which is obtained from the total system by integrating out the environmental degrees of freedom. It obeys a master equation. The local entropy increases during the emergence of entanglement with the environment. Again, decoherence only works if a special initial condition – a condition of weak entanglement – holds. But where can such a condition come from? We shall discuss a possible answer below.

The last of the main arrows is the gravitational arrow of time. Although the Einstein field equations are time-reversal invariant, gravitational systems in Nature distinguish a certain direction: the universe as a whole *expands*, while local systems such as stars form by *contraction*, for example from gas clouds. It is by this gravitational contraction that the high temperature gradients between stars such as the Sun and the empty space arise. Because of the negative heat capacity for gravitational systems, homogeneous states possess a low entropy, whereas inhomogeneous states possess a high entropy – just the opposite behaviour than for non-gravitational systems.

An extreme case of gravitational collapse is the formation of black holes. According to general relativity, their gravitational field is so strong that nothing, not even light, can escape. In spite of this, they possess fundamental thermodynamical properties (temperature and entropy) which are connected with their event horizon. The time-reversed versions of black holes, the so-called white holes, do not seem to exist in our universe. Temperature and entropy of a black hole exhibit their meaning if quantum theory is taken into account: according to the Hawking effect (see e.g. Kiefer (1999) for a detailed review), a black hole radiates with the Hawking temperature, $T_H$, which is proportional to the quantum of action $\hbar$. In the simplest case of a Schwarzschild black hole which is fully characterized by its mass, $M$, it reads

$$T_H = \frac{\hbar c^3}{8\pi G k_B M} \approx 6.2 \times 10^{-8} \frac{M_\odot}{M} \text{ K} \tag{4.2}$$

where $c$ is the speed of light, $G$ the gravitational constant, $k_B$ Boltzmann's constant, and $M_\odot$ the solar mass.

Connected with this temperature is the Bekenstein–Hawking entropy, $S_{BH}$, which is proportional to the area, $A$, of the black-hole horizon,

$$S_{BH} = k_B \frac{A c^3}{4 G \hbar} \stackrel{\text{Schwarzschild}}{\approx} 1.07 \times 10^{77} k_B \left(\frac{M}{M_\odot}\right)^2 \tag{4.3}$$

One can estimate from this entropy formula that the universe would be in a state of maximal entropy if all the observable matter were in the form of a single black hole (Penrose, 1990). Since the universe is far away from such a state, it must have been started with a very special initial condition. This would then be the desired boundary condition of low entropy mentioned above. Can this situation be analysed further? Close to the big bang, the classical theory of general relativity breaks down. It is most likely that a quantum theory of gravity must be invoked to describe this early phase, so to investigate this issue further one must address the field of quantum gravity.

## 4.2 Quantum gravity

There are many reasons why one should expect a quantum theory of gravity to supersede general relativity at the fundamental level (Kiefer, 2007). Perhaps the most important argument is the existence of singularity theorems indicating the breakdown of the classical theory at the big bang, the interior of black holes, and other regions (see below). Another important problem pointing towards an encompassing new theory is the *problem of time*: whereas time in ordinary quantum theory is non-dynamical, it *is* dynamical in general relativity, since it is part of spacetime obeying the Einstein equations. Both theories can thus not simultaneously be exactly correct.

The task of quantizing gravity has not yet been accomplished, but approaches exist within which sensible questions can be asked. Currently the two main approaches are: *superstring theory* (or *M-theory*) and *quantum general relativity*. Superstring theory is more ambitious and aims at a unification of all interactions within a single quantum framework. Quantum general relativity, on the other hand, attempts to construct a consistent, non-perturbative, quantum theory of the gravitational field on its own. This is done through the application of standard quantization rules (canonical methods, path integral methods, ...) to the general theory of relativity. A comprehensive review of these approaches can be found in Kiefer (2007).

In the following discussion I shall restrict myself to the canonical theory, and in particular to the version in which the metric plays the central role ('quantum geometrodynamics'). Another version of current interest is 'loop quantum gravity' in which the fundamental variable is a 'holonomy' that contains the integral of a 'vector potential' along a one-dimensional curve. While technically being more involved, the conceptual situation with respect to time is similar to quantum geometrodynamics (Rovelli, 2004).

What are the central equations of canonical quantum gravity? Since the classical theory (general relativity) possesses certain invariance properties (invariance with respect to coordinate transformations), it contains *constraints*. These are equations without second time derivatives. Because one has the freedom to choose four coordinates locally, there are four constraints at each space point. I shall not go into any details here (see Kiefer, 2007). The important fact is that the central equation, which is called the Wheeler–DeWitt equation, has the form of a zero-energy Schrödinger equation,

$$\hat{H}\Psi[h_{ab}, \varphi] = 0 \qquad (4.4)$$

where $\hat{H}$ denotes the total Hamilton operator of both gravitational and non-gravitational degrees of freedom. The gravitational part is represented by the *three*-dimensional metric, $h_{ab}$, and the non-gravitational part is, for simplicity, represented by a scalar field $\varphi$. The Wheeler–DeWitt equation has various characteristic features:

- it depends on the three-dimensional metric, but is invariant under coordinate transformations, so it depends, in effect, only on the three-dimensional geometry;
- it does not contain any time parameter and is thus of a 'timeless nature' at the most fundamental level;
- its general form follows from any theory which classically is time-reparametrization invariant;
- it is (locally) hyperbolic and thereby defines a concept of an 'intrinsic time' which has to be constructed from the spatial metric itself (and the matter degrees of freedom).

Spacetime as a classical concept has thus disappeared. How can this be understood? In classical canonical gravity, a spacetime can be represented as a 'trajectory' in configuration space – the space of all three-metrics. Although time coordinates have no intrinsic meaning in classical general relativity either, they can nevertheless be used to parametrize this trajectory in an essentially arbitrary way. Since no trajectories exist anymore in quantum theory, no spacetime exists at the most fundamental level, and therefore also no time coordinates to parametrize a trajectory.

In its most general form, (4.4) is not amenable to a mathematical treatment. For the following discussion it is, however, sufficient to turn to simple models. The Wheeler–DeWitt equation (4.4) can be greatly simplified if most of the degrees of freedom are frozen out already at the classical level and only few variables are quantized. A particular important example is the case of a homogeneous and isotropic universe. This is characterized classically by its scale factor ('radius') $a(t)$. If, in addition, a homogeneous scalar field, classically described by $\phi(t)$, is taken into account, one arrives effectively at a two-dimensional model. In quantum gravity, then, the wave function in (4.4) becomes a function of $a$ and $\phi$ and is denoted by $\psi(a, \phi)$. Recall that the classical time parameter $t$ has disappeared in the quantum theory, so no additional $t$ appears as an argument in the wave function.

If one has, for example, a massive scalar field with mass $m$ coupled to a closed Friedmann universe, (4.4) is of the form (in suitable units where $\hbar = c = G = 1$)

$$\left( a^2 \frac{\partial^2}{\partial a^2} + a \frac{\partial}{\partial a} - \frac{\partial^2}{\partial \phi^2} - a^4 + m^2 \phi^2 a^6 \right) \psi(a, \phi) = 0 \qquad (4.5)$$

One recognizes explicitly the hyperbolic ('wave') nature of this equation from the indefinite nature of the kinetic term. Since $t$ has disappeared, boundary conditions have to be formulated with respect to the configuration variables of the Wheeler–DeWitt equation, which are in this model given by $a$ and $\phi$. Since this equation is hyperbolic with respect to $a$,[1] $a$ can play the role of an 'intrinsic time' with respect to which boundary conditions can be imposed in a proper way.

This new concept of time has drastic consequences if one wants to describe a classically recollapsing universe in quantum gravity, see Kiefer (1988) and Kiefer and Zeh (1995). Both big bang and big crunch correspond to the *same* region in configuration space – the region of $a \to 0$. They are thus intrinsically indistinguishable. The Wheeler–DeWitt equation connects larger scale factors with smaller scale factors, but not the two ends of a classical trajectory (which is absent in the quantum theory). If one wants to mimic the classical trajectory by a 'recollapsing' wave packet, one has to include both the 'initial' and the 'final' wave packet into one initial condition with respect to $a$. If one of the two packets were lacking, one would not be able to recover the classical trajectory as an approximation. It also turns out that it is in general not possible to construct a narrow wave tube all along the trajectory corresponding to a recollapsing universe – a dispersion in certain regions is unavoidable. Therefore, a semiclassical approximation is not valid all along the trajectory and quantum effects can play a role even far away from the Planck scale – for example at the turning point of the classical universe.

Macroscopic quantum effects in cosmology can also occur in models that classically contain a singularity for a *large* universe. Such singularities can be a 'big rip', in which the universe expands to an infinite size in a finite time, or a 'big brake' in which the universe comes to an abrupt halt (with infinite deceleration) in the future. The corresponding quantum theories have been discussed in Dąbrowski *et al.* (2006) ('big rip') and in Kamenshchik *et al.* (2007) ('big brake'). It turns out that they can be formulated in a singularity-free way. However, quantum effects turn over at the place of the classical singularities, which means that the classical evolution comes to an end and time, together will all classical observers, stops to exist, too.

If one, alternatively, uses loop quantum gravity as the starting point, one will arrive at the framework of loop quantum cosmology, see, for example, Bojowald (2006). Near the Planck scale, the Wheeler–DeWitt equation becomes a difference equation that could entail the possibility of avoiding the classical singularity in the quantum theory. The full consequences of this approach for the concept of time remain to be explored.

---

[1] This becomes evident if further degrees of freedom are added; they all occur with the same sign as $\phi$ in the kinetic term.

### 4.3 Low-entropy condition from quantum gravity

As we saw in the previous section, the fundamental equation of canonical quantum gravity does not contain any time parameter. The same holds for string theory. How, then, can one derive a direction of time if there is no time in the first place?

The answer involves two steps. The first step consists in the recovery of an *approximate* time parameter from the timeless Wheeler–DeWitt equation (4.4). This is achieved through a Born–Oppenheimer type of approximation scheme (Kiefer, 2007). If one makes for particular solutions of (4.4) the ansatz

$$\Psi[h_{ab}, \varphi] = e^{iS[h_{ab}]/G} \Phi[h_{ab}, \varphi] \tag{4.6}$$

it turns out that to highest order in an expansion with respect to the gravitational constant, the functional $S[h_{ab}]$ obeys the gravitational Hamilton–Jacobi equations, which are equivalent to the classical Einstein equations. Solutions of these equations *define* a semiclassical background spacetime. The wave functional $\Phi[h_{ab}, \varphi]$ then obeys – in the next order in $G$ – a functional Schrödinger equation in which the time $t$ is *defined* through this approximate background,

$$i\hbar \frac{\partial}{\partial t} \Phi(t, \varphi] = \hat{H}^{\mathrm{mat}} \Phi(t, \varphi] \tag{4.7}$$

where $\hat{H}^{\mathrm{mat}}$ denotes the non-gravitational part of the total Hamiltonian. This is the limit of quantum field theory in an external spacetime. In cosmological models (cf. the previous section) the semiclassical time $t$ is usually correlated with the scale factor $a$.

The second step in the search for an arrow of time seeks an answer to the question: how can a *direction* of this semiclassical time emerge? It is a property of the full Wheeler–DeWitt equation – and already recognizable in the simple model (4.5) – that the potential term is *asymmetric* with respect to intrinsic time. In fact, the potential becomes very simple for $a \to 0$, approximately separating between all variables. One can thus impose an initial condition of lacking entanglement between the various degrees of freedom, that is, one can impose the condition

$$\Psi \xrightarrow{a \to 0} f(a) \prod_i \chi_i(x_i) \tag{4.8}$$

where $x_i$ is a formal notation for all variables except $a$ and the $\chi_i$ are the corresponding wave functions (Zeh, 2007). With this initial condition near the 'big bang', (4.4) yields an increasing entanglement for increasing $a$ because the potential term couples the various degrees of freedom. For subsystems, this then leads to an increase of entropy and to increasing decoherence. This increase constitutes the origin of both the thermodynamical arrow (and, consequently, also the radiation arrow) and the quantum-mechanical arrow of time. In fact, it becomes clear that all arrows are then correlated with the radius of the universe, and that the cosmological arrow plays the role of the master arrow of time, as expected. The universe can then *only* be observed as expanding (Kiefer and Zeh, 1995).

As long as the final theory of quantum theory remains elusive, these considerations are, of course, speculative. They show, however, how one can hope to understand *at least in principle* the origin of irreversibility from a fundamental theory of Nature.

## References

Bojowald, M. (2006). Universe scenarios from loop quantum cosmology. *Annalen der Physik*, **15**, 326–341.

Dąbrowski, M. P., Kiefer, C. and Sandhöfer, B. (2006). Quantum phantom cosmology. *Physical Review D*, **74**, article number 044022.

Joos, E., Zeh, H. D., Kiefer, C., Giulini, D., Kupsch, J. and Stamatescu, I.-O. (2003). *Decoherence and the Appearance of a Classical World in Quantum Theory*, 2nd edn. Berlin: Springer-Verlag. See also http://www.decoherence.de.

Kamenshchik, A. Y., Kiefer, C., and Sandhöfer, B. (2007). Quantum cosmology with a big-brake singularity. *Physical Review D*, **76**, 064032.

Kiefer, C. (1988). Wave packets in minisuperspace. *Physical Review D*, **38**, 1761–1772.

Kiefer, C. (1999). Thermodynamics of black holes and Hawking radiation. In *Classical and Quantum Black Holes*, ed. P. Fré *et al.* Bristol: IOP Publishing, pp. 17–74.

Kiefer, C. (2007). *Quantum Gravity*, 2nd edn. Oxford: Oxford University Press.

Kiefer, C. and Zeh, H. D. (1995). Arrow of time in a recollapsing quantum universe. *Physical Review D*, **51**, 4145–4153.

Penrose, R. (1990). *The Emperor's New Mind*. Oxford: Oxford University Press.

Rovelli, C. (2004). *Quantum Gravity*. Cambridge: Cambridge University Press.

Zeh, H. D. (2007). *The Physical Basis of the Direction of Time*, 5th cdn. Berlin: Springer-Verlag. See also http://www.time-direction.de.

# Part II

Probability and chance

# 5

# The natural-range conception of probability

JACOB ROSENTHAL

## 5.1 Objective interpretations of probability

Objective interpretations claim that probability statements are made true or false by physical reality, and not by our state of mind or information. The task is to provide truth conditions for probability statements that are objective in this sense. Usually, two varieties of such interpretations are distinguished and discussed: frequency interpretations and propensity interpretations. Both face considerable problems,[1] the most serious of which I will briefly recall to motivate the search for an alternative.

Firstly, the frequency interpretations. Here the central problem is that it is very implausible (to say the least) to postulate a non-probabilistic connection between probabilities and relative frequencies. What a frequency approach claims seems either to be false or to presuppose the notion of probability. Take, for example, the repeated throwing of a fair die that has equal probabilities for each side. All you can say is that it is *very probable* that upon many repetitions each face will turn up with a relative frequency of approximately 1/6 (weak law of large numbers). Or that, *with probability* 1, the limiting relative frequency of each face would be 1/6 in an infinite number of repetitions (strong law of large numbers). You cannot drop the clauses 'very probable' or 'with probability 1' in these statements. There are no relative frequencies that the die *would* produce on repeated throwing, but it *could*, with varying probabilities, yield any frequency of a given outcome. This is not only clear from the standard mathematical treatment, according to which the throws are probabilistically independent, but one should expect it from the outset. After all, a chance experiment is repeated, so it should be possible to get the same

---

[1] Hájek (1997, 2009) gives a comprehensive list of those problems for frequency accounts, Eagle (2004) does the same for propensity accounts of probability.

result over and over again.[2] The idea of repeating a random experiment does not fit together with the claim that certain relative frequencies would or have to emerge. So, you cannot use the connection between probabilities and relative frequencies for an interpretation of the former. Any such connection is itself probabilistic and given by the laws of large numbers, which cannot plausibly be strengthened. Nor will it do to give a different interpretation to those second-order probabilities. They arise from a combination of the first-order probabilities, and thus there is no reason for interpreting them in another way.

Second, in the so-called single-case propensity theories, the entities called 'propensities' are supposed to be fundamental entities in Nature which somehow attach to individual physical processes and make their different possible outcomes more or less probable. So, propensities are, or create, irreducible objective single-case probabilities. But this implies that they constrain rational credence, because otherwise the term 'probability' would be misplaced. Propensities yield the appropriate degrees of belief. There is no account of how they might be able to achieve this, they constrain rational credence 'just so'. They are assumed to be normative parts of reality, and their normativity is to be taken as a brute fact. If one tries to avoid this oddity by explicating propensities in terms of relative frequencies, one is saddled again with the central problem of the frequency approach. The so-called long-run propensity theories are actually varieties of the frequency interpretation and share most of its features and problems, whereas the single-case approaches do not have much to say about probabilities. They just say that there are entities out there which somehow adhere to single events and give us appropriate degrees of belief. This austere fact is concealed by the introduction of various new notions, like 'disposition of a certain strength', 'weighted physical possibility', 'causal link of a certain strength' or even 'generalized force'. 'Propensity' is one of these terms, too, which are in greater need of explication than 'probability' itself, and the explication of which inevitably leads back to the very concept of probability they are supposed to explain.

In view of these difficulties it does not seem superfluous to look for yet another objective interpretation of probability. And in fact there is a third, not actually new but to a large extent neglected, possibility for such an interpretation: namely, probabilities as deriving from ranges in suitably structured spaces of initial states. To put it very briefly, the probability of an event is the proportion of those initial states within the initial-state space attached to the underlying random experiment which lead to the event in question, provided that the space has a certain structure.

---

[2] As usual in discussions of probability, I use the term (chance or random) 'experiment' and related notions like 'experimental arrangement' in a wide sense throughout, standing for 'in principle repeatable physical process or constellation'. It is not just to be thought of experiments that are or can be conducted by human experimentators.

I will call this the 'range interpretation' of probability statements. It was already proposed by Johannes von Kries, by the end of the nineteenth century, in his comprehensive book *Die Principien der Wahrscheinlichkeitsrechnung* (von Kries, 1886, 2nd printing 1927). Where I will use the term 'range', von Kries spoke about 'Spielraum', which could also be translated as 'leeway', 'room to move', or 'play space'.[3] I will present this interpretation by making a brief detour through the classical conception of probability, to which it is related.

## 5.2 From the classical interpretation to the range interpretation

According to the well-known classical conception (CC), the probability of an event is the ratio of 'favourable' (with regard to the event in question) to 'possible' cases, where those possible cases must be 'equally possible' and are judged so by the 'principle of indifference' or 'principle of insufficient reason':

(CC) Let $E$ be a random experiment. On every trial of $E$ exactly one of the cases $a_1, a_2, \ldots, a_n$ is realized. If we do not see (subjectivist reading) or there is not (objectivist reading) any reason why $a_i$ should occur rather than $a_j$ for each $i, j$, then the cases $a_1, a_2, \ldots, a_n$ are equally possible. If exactly $k$ of them lead to the event $A$, then the probability of $A$ on a trial of $E$ is $k/n$.

This classical conception, which originated with the invention of the calculus of probabilities by Fermat and Pascal in the seventeenth century and culminated with Laplace, was abandoned in the course of the nineteenth century due to the following difficulties: first, equally possible cases are not always to be had, i.e. the principle of insufficient reason is not always applicable. Second, the opposite problem: sometimes there are different types of equally possible cases, i.e. the principle of insufficient reason is applicable in different ways which lead to contradicting probability ascriptions (Bertrand's paradoxes). Third, since 'equally possible' can mean nothing but 'equally probable', the classical interpretation provides no analysis of probability, but presupposes the concept. As will become clear, the range interpretation is able to overcome the first and the second problem. The third one proves harder and will turn out to be a touchstone for success or failure of the range approach.

Now, for example, in how far are the six sides of a symmetric die 'equally possible' upon throwing? In how far *is* there *not* any reason (remember we are interested in an *objectivist* reading of the principle of insufficient reason) why one side should turn up rather than another? In how far is it equally easy to get a certain number $a$ as the outcome and to get any other number? Take the outcome

---

[3] See Heidelberger (2001).

of a throw to be determined by its initial conditions. These can be represented by a vector of real numbers, which characterize things like the position of the die in space, its velocities in the different directions, its angular velocities in the different directions, etc., at the very beginning of its flight. Call any such vector, which represents a complete selection of initial conditions, an initial state. Thus, for some $n$, the $n$-dimensional real vector space $\mathbf{R}^n$, or a suitable subspace of it, can be viewed as the space of possible initial states for a throw of the die. Each point in the initial-state space $S$ fixes the initial conditions completely, and thus fixes a certain outcome of the throw. Here it is assumed that other conditions, which one could call 'boundary conditions', remain constant. These include the physical properties of the die and of the surface onto which it is thrown, the air pressure and temperature and other things like that. The difference between initial and boundary conditions is made in order to distinguish varying factors from constant ones. If, e.g., the air pressure might change from one throw to the next, it would also have to be included among the initial conditions. Strictly speaking, we would then face a different random experiment with a different initial-state space, but, at least in ordinary cases, with the same outcome probabilities.

If we could survey the initial-state space, we would find it consisting of very small patches where all the initial states contained lead to the same outcome. In the neighbourhood of a patch associated with the number $a$ as outcome we would find patches of approximately equal size leading to any other number. Therefore, the outcome $a$ is represented with a proportion of approximately 1/6 in any not-too-small segment of $S$. Without loss of generality, we can lay down this condition for intervals: Let $S_a \subseteq S$ be the set of those initial states that lead to outcome $a$. Let $I \subseteq S$ be an interval. Let $\mu$ be the standard (Lebesgue-) measure. If $I$ is not too small, we will find that

$$\frac{\mu(\mathbf{I} \cap \mathbf{S}_a)}{\mu(\mathbf{I})} \approx 1/6$$

The proportion of initial states that lead to outcome $a$ in an arbitrary, not-too-small interval $I$ of the initial-state space is about 1/6.

A loaded die can obviously be treated in exactly the same manner, i.e. the approximate probabilities attached to its sides can be found in any not-too-small segment of the space of initial states as the proportion of those initial states in that segment which lead to the respective outcome. This already overcomes the first shortcoming of the classical interpretation. The structure of the initial-state space explains the typical probabilistic patterns that emerge on repeated throwing of the die: why it is impossible to predict the outcome of a single throw, on the one hand, and why the different outcomes occur with certain characteristic frequencies in the long run, on the other hand. Of course, I have not *shown* that the initial-state space associated with the throwing of a certain die has the sketched structure. This would be a very demanding task. I have to assume that it is possible in principle to do so, using only classical mechanics, and hope that the reader finds this assumption plausible.

## 5.3 The natural-range interpretation of probability

The basic idea underlying the range conception is as follows: When we observe a random experiment that gives rise, on repetition, to probabilistic patterns, it is often plausible to suppose, as I did with the die, that this phenomenon can be attributed to an initial-state space with a structure of the indicated kind. On the one hand, in any (not too) small vicinity of an initial state leading to a given outcome $a$ there are initial states leading to different outcomes, which explains why the outcome of a single trial cannot be predicted. On the other hand, for each outcome, the proportion of initial states leading to it is constant all over the initial-state space, i.e. it is approximately the same in any not-too-small segment of the space, which explains why there are certain stable characteristic relative frequencies with which the different outcomes occur. We cannot use those frequencies to define the probabilities of the respective outcomes – remember the main objection to frequency interpretations – but we can use the underlying initial-state space to do so.

Before setting up the definition, however, we have to specify the requirement that a certain outcome $a$ is represented with approximately the same proportion in any not-too-small *segment* of the initial-state space. This proves a bit tricky, because if we demand too much here, the applicability of the range interpretation will be unnecessarily restricted, whereas if we demand too little, certain mathematically desirable features of the range approach are lost. First, we have to presuppose that the initial-state space $S$, as well as the set $S_a \subseteq S$ of those initial states that lead to the outcome $a$, are Lebesgue-measurable subsets of $\mathbf{R}^n$ (for some $n$) of maybe infinite size. More precisely (to avoid problems with infinity), we assume that the intersection of $S$ or $S_a$ with any bounded measurable subset of $\mathbf{R}^n$ is again measurable. Second, we look at those measurable subsets of $S$ whose size exceeds a certain minimum; let us call them 'not too small' from now on. We certainly cannot assume that $a$ is approximately equally represented in any such set, because $S_a$ is itself one of these sets. A natural idea is to stick to bounded and connected measurable sets, or *areas*. As any area can be approximated by ($n$-dimensional) intervals, we could, without loss of generality, set up the proportion-condition for *intervals*. Unfortunately, even this proves too strong. Imagine a random experiment with a two-dimensional initial-state space, such that the 'patches' the space consists of are extremely narrow but potentially infinitely long stripes. As the stripes are so narrow, the range conception of probability should be applicable, but as they are at the same time very long, there are quite large intervals within which all initial states give rise to the same outcome. So it seems that we ought to be even more restrictive and consider only *cube-shaped* or *equilateral intervals*. This is somewhat inelegant, but in this way we are able to capture exactly those cases in which the range approach is intuitively applicable, without giving up too much of the generality of the proportion-condition. Now we are in a position to state the range conception (or interpretation, or definition) of probability, (RC).

(RC) Let $\boldsymbol{E}$ be a random experiment and $a$ a possible outcome of it. Let $\boldsymbol{S} \subseteq \mathbf{R}^n$ for some $n$ be the initial-state space attached to $\boldsymbol{E}$, and $\boldsymbol{S}_a \subseteq \boldsymbol{S}$ be the set of those initial states that lead to the outcome $a$. Let $\mu$ be the standard (Lebesgue-) measure. If there is a number $p$ such that for any not-too-small $n$-dimensional equilateral interval $\boldsymbol{I} \subseteq \boldsymbol{S}$, we have

$$\frac{\mu(\mathbf{I} \cap \boldsymbol{S}_a)}{\mu(\mathbf{I})} \approx p$$

then there is an objective probability of $a$ upon a trial of $\boldsymbol{E}$, and its value is $p$.

This is the first formulation of the range conception. It would be more accurate to speak of the 'physical-' or 'natural-range conception' (see comment K below), but I will stick to the short term. A second formulation of this approach to probabilities will be given in the next section, using so-called 'arbitrary functions'. Before turning to that, I am going to make several comments on (RC).

(A) The range approach is meant to be a proposal for an objective interpretation of probability statements, i.e. as providing truth conditions for such statements that do not depend on our state of mind or our state of information and are therefore called 'objective'. The interpretation is reductive: probability is analysed in non-probabilistic terms. The axioms of the calculus of probability immediately follow from the definition, if one takes a small idealizing step and reads the approximate equation as an exact one. The central problem of frequency interpretations is avoided, because the relative frequency of a given outcome may, upon repetition of the random experiment, deviate as much as you like from the proportion with which the outcome is represented within the initial-state space. Such an event is not excluded, merely very improbable. Many of the problems of frequency approaches can be traced back to the confusion of assertibility conditions with truth conditions. What makes a statement about an objective probability true is not the frequencies that will or would arise or have arisen, but what explains those frequencies – in a probabilistic way, using the laws of large numbers. The statement is made true by the physical circumstances that give rise to the typical probabilistic pattern. The range approach claims to spell out what those circumstances are.

(B) Second-order probabilities of the kind just mentioned, as they occur in the laws of large numbers, can also be given a range interpretation in a natural way. Considering $k$ independent repetitions of $E$ as a new, complex random experiment $E^k$, the initial-state space attached to this new experiment is just the $k$-fold Cartesian product – call it $S^k$ – of $S$ with itself. The probabilities of the possible outcomes of $E^k$ are the proportions with which these outcomes are represented in $S^k$. Therefore, if the probabilities in a random experiment admit of a range interpretation, so do the probabilities associated with independent repetitions of the experiment, and the respective ranges are connected in a straightforward way. I will not discuss the problems connected with the notion of independence, in particular the very important question of how to justify the stochastic independence of different repetitions of $E$, but refer the reader to Strevens (2003: ch. 3). Suffice it to say that the independence of the range probabilities need not be taken for granted, or simply assumed, but can be explained in much the same way as the existence of the probabilities themselves. It is a virtue of the approach that the question concerning the source of the objective probabilities and the one concerning the source of their independence on repeated trials admit of largely parallel answers.

(C) The range interpretation works only in deterministic contexts, where the outcome of a chancy process is determined by initial conditions.[4] This means that terms like 'random experiment' or 'chance process' cannot be taken in a hard realist sense. What is random or chancy in this sense depends on our epistemic and computational abilities. So, according to the range interpretation, *that* there are probabilities depends on us, but *what* they are depends on the world. The structure of an initial-state space of the required kind, and in particular the proportions with which the various outcomes are represented within it, is perfectly objective. It only depends on the laws of Nature and the dynamics of the process in question, but not on our state of mind or information. But that we call those proportions 'probabilities', and call the underlying experiment a 'random' or 'chance experiment', is dependent on the epistemic and control capacities we possess (or are willing to exercise). There is an epistemic aspect to range probabilities, but this aspect does not concern their value. It concerns the fact that certain experiments are considered to be random or chancy.

---

[4] One could try to generalize the approach by considering processes in which an initial state does not fix an outcome, but only a probability distribution over the possible outcomes. But this distribution would have to be interpreted in turn, and if this were done according to the range approach, one could in principle eliminate these probability distributions by moving to another, higher-dimensional initial-state space. Therefore, the range approach is indeed restricted to deterministic contexts, or else becomes dependent on a different interpretation of probability.

(D) There is a reservation to the range conception that may be addressed in connection with the example of a loaded die. One might speculate that with a die which is, e.g., biased in favour of 'six', we do not get the same proportion of 'six' all over the initial-state space, but that this proportion increases when the die is thrown higher above the surface. If anything of this kind were true, there would neither be a characteristic relative frequency of 'six' to be explained, nor an objective probability of this outcome. We would still face a random experiment with an initial-state space, to be sure, but we could not attach objective probabilities to the possible outcomes – at least not in the usual sense of point values. To get objective probabilities proper, we would have to constrain the method of throwing by demanding that the die is thrown in a certain fixed distance to the surface. What *could* rightly be said, however, is that the objective probability of 'six' in the original random experiment is greater than 1/6. I do not think that suchlike occurs with a loaded die, but random experiments of this kind certainly exist. In these cases, the range approach does not provide definite, point-valued objective probabilities – even abstracting from the fact that (RC) merely states an approximate equality – nor are there any to be had. We may, however, get intervals for objective probabilities. It should not be too surprising that there may be random experiments which in this way give rise to interval-valued objective probabilities.

(E) The definition (RC) is vague in two respects: The proportion of initial states leading to outcome $a$ is *approximately* equal to $p$ in any *not-too-small* cubic interval of the initial-state space $S$. This is somewhat unsatisfactory; the first vagueness, in particular, has the consequence that range probabilities do not have exact point values. These are rather chosen for mathematical convenience. The ideal case would be the truly chaotic limiting case in which for *every* equilateral interval $I$ we have $\dfrac{\mu(I \cap S_a)}{\mu(I)} = p$ *exactly*. This never occurs in reality, but often it is convenient to model chancy situations as if this were true. We approach the limiting case, for example, if in the throwing of a die we consider only throws that are made high above the surface with high initial angular velocities. That a situation appears chancy to us at all is due to the fact that it approximates the limiting case sufficiently; how good the approximation must be depends again on our epistemic and control abilities. They determine what the clause 'not too small' in (RC) amounts to, i.e. what is the minimal interval size from which onward the condition is required to hold.

(F) The range definition of probability stands in-between two more clear-cut proposals. First, just take the proportion of an outcome in the initial-state space as the probability of this outcome. Do not care how the initial states associated with this outcome are distributed over the space. Second, only in the truly chaotic limiting case do we have objective probabilities. The equation has to hold *exactly* for *every* cube-shaped interval. The first proposal is untenable. If the objective probability of an event were just its proportion in the initial-state space, no matter how this space is structured, we would have to ascribe objective probabilities in cases that are not at all random to us, while in other cases we would get the wrong probabilities out of this proposal. This is because on repetition of the random experiment, we cannot expect the initial states that actually occur to be evenly distributed over the initial-state space. There will be regions of $S$ that occur more often than others, and we have to guarantee that this does not influence the probabilities. Therefore, we have to require that the proportion of the initial states leading to a given outcome is (about) the same in any (not-too-small) segment of $S$.

The second proposal is reasonable. One may constrain talk of objective probabilities to the limiting case. This has two advantages: it is impossible *in principle* to influence the probabilities by refinement of our abilities of measurement, computation and control. Nature may know the outcome in advance, but we will never know, nor can we improve in any way on the value $p$ as an expectation whether or not the outcome $a$ will occur on a single trial. Furthermore, the outcome probabilities are given precise numerical values by the structure of the initial-state space. The disadvantage, of course, is that this kind of situation never occurs in reality. But here one may simply say that many actual situations are very close to the limiting case, i.e. practically indistinguishable from it for us, given our epistemic capacities. Therefore we model them accordingly. As reality contains structures that can play the role of objective probability near enough, the ascription of objective probabilities is justified in such cases.

(G) We do not get single-case probabilities out of the range approach. The initial-state space is something attached to a *type* of random experiment, whereas in a single trial, there is a definite initial state as well as a definite outcome, and no more. We neither have several possible outcomes, one of which is realized, nor an initial-state space, one element of which occurs – after all, a deterministic context is presupposed throughout. I have used the term 'random experiment' in the type-sense, as I spoke about repetitions of a given experiment, and without a notion of the repetition or, equivalently, of the type of a given process, the talk of possible outcomes or possible initial states makes no sense. Consequently there is no way to get anything like range probabilities for a single case considered in isolation. According to the range approach an objective probability is never single-case, and a single-case probability requires an epistemic reading. This is not surprising, because objective single-case probabilities imply genuine indeterminism. It may be tempting to regard as indetermined which initial state is realized on a given occasion and to get objective single-case probabilities in this way, after all. That would also cut off the problem discussed in the next remark. But this move would make the occurrence of a particular initial state the outcome of a further, now genuinely indeterministic random experiment, and consequently suggest a further concept of probability. In order to preserve the point of the range approach, one has at least to remain agnostic regarding the question whether or not the occurrence of a certain initial state is itself determined.

(H) The term 'random experiment' is used in a broad sense, meaning 'in-principle-repeatable physical process', and exhibiting no particular connection to epistemic subjects that are thought to 'intervene in Nature'. It is therefore a matter of convention what one picks out as the initial state of a certain process. Take again the throwing of a die. The outcome is determined by initial conditions, but why take as initial conditions various physical parameters at the moment when the die leaves the gambler's hand? Why not go further back in time and look, e.g., at the moment when he takes up the die? It seems that one could as well regard the physical circumstances at that moment, which include a physical characterization of the gambler's brain and body, as the initial conditions of the throw. Obviously, there are in principle countless possibilities for what to take as the initial-state space of the experiment 'throw of a die with such-and-such properties under normal circumstances'. Which one gives us the true probabilities?

In reply to this, one may simply say that these cases amount to several different random experiments with different initial-state spaces. In specifying a type of experiment or a type of physical process, one has to specify in particular what counts as the beginning of the process or experiment. That cuts off the problem, but at the cost of threatening to completely undermine the objectivity of probability ascriptions to event types like 'getting six on a throw of a symmetric die under normal circumstances'. One can give a more substantial answer by observing that going further back in time presumably would not change the probabilities, and making such a feature a *requirement* for probability ascriptions according to the range conception. This is von Kries's line (1886: ch. 2.3, 2.6). According to him, the ranges in an initial-state space provide objective probabilities only if they are 'ursprünglich' (primordial, original), which means just that: if one regards earlier states as the initial states, this does not affect the respective proportions of the outcomes of the process. To be sure, one gets many different 'initial'-state spaces for a given type of process, but always with the same properties with regard to the possible outcomes. Von Kries also provides arguments to the effect that normally we are justified in regarding the state spaces we are inclined to view as initial as giving rise to primordial ranges.

I leave open which route to take here, since the problem leads into very deep waters. The easy solution implies that range probabilities are relativized in the mentioned respect, and it thus becomes difficult to uphold the claim that they are objective. They depend on a cut in time that is conventionally made by epistemic subjects. The ambitious solution, on the other hand, seems to imply that whenever we ascribe objective probabilities to certain events, we have to make far-reaching assumptions. We have to be able to locate the 'initial state' of the process – and with it the relevant state space – arbitrarily far in the past without changing the probabilities. Even if we are ready to make such assumptions, however, at some stage we will hit on the initial state of the universe, and it is doubtful whether talk about an 'in-principle-repeatable physical process' still makes sense then. At this stage, the 'repetition of the experiment' would amount to the creation of a new universe with the same laws of Nature but, presumably, a different initial state out of the space of all possible initial states of the universe. And considering *this* state space, which no doubt rightly carries the name 'primordial', the proportions of the outcomes would *still* have to be the same. It would be quite embarrassing if one were (implicitly) committed to such considerations when ascribing range probabilities. It may however be the price for full objectivity, and if we are not prepared to pay it, we may have to be content with a pragmatically reduced range concept of probability.

(I) In what sense is the range interpretation related to the classical one? One could say that the former treats the initial states in any small equilateral interval of $S$ implicitly as 'equally possible', because it identifies the probability of a certain outcome $a$ with the proportion with which $a$ appears in such an interval. One could interpret this as the implicit supposition of an approximately uniform distribution over any small cubic interval in $S$. Therefore, the 'equally possible' cases are in a weak sense always to be had, after all: initial states that are near to one another in $S$ may be said to be treated implicitly as equally possible by (RC). This already points to the main problem of the range approach, to be discussed in Section 5.5: as with the classical conception, 'equally possible' means nothing but 'equally probable' here, so the concept of probability seems to be presupposed by the range approach. But the approach is applicable to loaded dice and the like. Furthermore, it does not depend on a subjectively interpreted principle of indifference and consequently does not fall prey to Bertrand's paradoxes. So it can at least overcome the first two problems of the classical concept. Whether it is able to deal with the third problem, too, remains to be seen.

(J) It is also instructive to compare the range interpretation to frequency and propensity accounts of objective probability. As I said in remark A, frequency approaches mistake the evidence for objective probabilities for the probabilities themselves. No doubt there is a close connection between objective probabilities and frequencies: the former can be used to explain and predict the latter, the latter to infer the former. But all these explanations, predictions and inferences are probabilistic in themselves, and what makes an objective probability statement true is not the observed or expected frequency pattern, but the physical features of the experimental set-up that give rise to this pattern (in a probabilistic way). Now, it would be quite natural to *define* objective probability by saying simply this: the objective probability of an event is that very feature, whatever it may be, of the experimental set-up that brings about the characteristic probabilistic or frequency patterns. What Popper and others say concerning propensities *may* be understood along these lines, which would turn the propensity conception into a very general, but also nearly empty account of objective probability. Understood in this way, it is not subject to the above-mentioned criticisms, but it is also uninformative: it says nothing more than that probability statements refer to such properties of experimental arrangements that can play the role objective probabilities are supposed to play. But *what* can play this role, what features of experimental set-ups can rightly be called 'objective probabilities'? We do not yet have an interpretation of probability, but something more abstract. Given this reading, the propensity conception is, as it were, merely a promissory note that has yet to be cashed out. The range interpretation can be seen as obtaining this; it makes a proposal what those features of random experiments are. It is a reasonable conjecture, put forward by Strevens (2003: ch. 5), that it is precisely such constellations in which probabilistic patterns emerge and in which we can ascribe objective probabilities to certain events. Other proposals are made by the different propensity conceptions (this term now understood in the specific sense in which it was discussed at the beginning), notably by single-case theories. Whereas the range approach takes those features of experimental arrangements which can be identified as probabilities to be complex high-level properties, single-case propensity theories take them to be simple, irreducible and fundamental.

(K) As is clear by now, the 'ranges' we deal with here are physical or natural ranges. All examples in which the initial-state space can actually be surveyed come from physics, but in principle the approach is suited to all empirical sciences, a point argued for already by von Kries and again by Michael Strevens. What one needs to apply the range approach is a distinction between initial conditions and laws. Moreover, the initial conditions have to be viewed as forming a real vector space: as will become even clearer in the next section, we are in the realm of real analysis here. It may be possible to generalize the approach to more abstract topological (vector) spaces, but a notion of continuity is indispensable. I emphasize this, because von Kries's ideas received a rather one-sided reception. They were taken up by John Maynard Keynes as well as by the Vienna Circle, notably Friedrich Waismann and Ludwig Wittgenstein, who both used the term 'Spielraum', but gave it a distinctly logical meaning.[5] Since then, the term 'range', when it occurs in philosophical discussions of probability, usually refers to logical probabilities. But these face the same problems as classical ones and are supposed to be *a priori* and uniquely singled out by requirements of rationality. As is clear by now from the heroic attempts of Rudolf Carnap, there is nothing of this kind to be had. One gives away the potential of von Kries's ideas if one develops them only in the direction of a logical concept of probability. To avoid misunderstandings (RC) could be dubbed 'physical-range conception', if this were not too narrow. Again, what is essential are laws of Nature and continuous initial conditions, but not necessarily laws of physics. Therefore, the term 'natural-range conception' seems to be most appropriate.

## 5.4 The method of arbitrary functions

Continuous functions can be characterized as functions that are approximately constant on any sufficiently small interval. It is enough for continuity if this holds for equilateral intervals. Hence, in the ideal limiting case of the range interpretation we have

$$\int_{S_a} \delta(\mathbf{x}) \, d\mathbf{x} = p$$

for any continuous density function $\delta : \mathbf{S} \to \mathbf{R}$ from the initial-state space $\mathbf{S}$ into the real numbers $\mathbf{R}$. In reality, at best we approach the ideal case, and therefore the equation is neither exactly true nor does it hold for certain eccentric density functions. Two types of (continuous) densities are obviously apt to create problems: on the one hand, density functions that are periodic with a period exactly the size of the 'patches' the initial-state space is divided into, and on the other hand, density functions that put most of the weight on a single patch (a patch being a measurable connected set of maximum size consisting of initial states that all lead to the same outcome). These densities fluctuate either with a high frequency, or else very strongly, on at least one relatively small area of the initial-state space. In order to exclude them, we have to restrict the set of admissible densities to those of

---

[5] See von Wright (1982) and Heidelberger (2001).

appropriately bounded variation. Taking this into account, we can restate the range interpretation of probability thus:

(AF) Let $E$ be a random experiment and $a$ a possible outcome of $E$. Let $S \subseteq \mathbf{R}^n$ for some $n$ be the initial-state space attached to $E$, and $S_a \subseteq S$ be the set of those initial states that lead to outcome $a$. If there is a number $p$ such that for any continuous density function $\delta: S \to \mathrm{R}$ whose variation is appropriately bounded on small intervals of $S$, we have $\int_{S_a} \delta(\boldsymbol{x})\,d\boldsymbol{x} \approx p$, then there is an objective probability of $a$ upon a trial of $E$, and its value is $p$.

The bounded-variation-condition would have to be stated more precisely as follows: There must be appropriate positive real-valued constants $c$, $k$ such that for any continuous density function the variation of which is less than $c$ on any interval of a size less than $k$, the approximate equation holds. In this way densities are excluded that fluctuate either with a very high frequency or very abruptly and strongly. Let us call such densities 'irregular' or 'eccentric', the others 'regular', 'ordinary', or 'well-behaved'. The central question for the range approach, to be discussed in the next section, concerns the justification of this constraint. Before addressing it, I will continue the list of remarks.

(L) The approach to probabilities via continuous density functions is traditionally called 'method of arbitrary functions'. A sketch of it can be found in von Kries (1886: ch. 3). It is worth pointing out that von Kries addressed almost any aspect of the range conception, including, for example, its application to statistical physics (ch. 8). One can say with little exaggeration that the very first treatment of the approach was also essentially complete concerning the main ideas, but much less concerning their elaboration. Von Kries's style is rather informal, even more than that of the present chapter, which is partly due to the fact that the requisite mathematical tools were not yet developed in his times. Henri Poincaré, in *Calcul des Probabilités* (1896) as well as in *Science and Hypothesis* (1902), was the first to apply the method of arbitrary functions rigorously, to the so-called wheel of fortune. This wheel consists of a disc which is divided into small segments of equal size whose colours alternate between red and black, and a spinning pointer which is pushed and eventually comes to rest on either a red or a black segment. It is a particularly simple example, because, provided that the pointer starts in a fixed initial position, its final position depends on its initial velocity only, and so the initial-state space has only one dimension. Poincaré shows that any ordinary distribution over initial velocities leads to a probability of approximately 1/2 for the outcomes 'red' and 'black', respectively. Since then, the wheel of fortune has served as the standard example to introduce the method of arbitrary functions.

(M) Later on, more demanding examples were treated by Eberhard Hopf (1934, 1936), who also gave the first systematic account of the method. The difficulty is, not just to make a more or less plausible *claim* to the effect that with a certain arrangement the assumptions of the range interpretation or the method of arbitrary functions are fulfilled and that, therefore, probabilities that are estimated, or known in advance from other sources, can be understood accordingly, but to *show* that those assumptions are fulfilled and to *derive* the probabilities mathematically by actually constructing and investigating the initial-state space. This proves too difficult even in comparatively simple examples, like the Galton-board, but Hopf treated some non-trivial examples at least, in which the probabilities are not known in advance from symmetry considerations. In general, the range conception is meant to provide *truth* conditions for probability statements, but it is not well suited to supply *assertibility* conditions. In most cases, one has to infer the probabilities from symmetries or relative frequencies instead of directly assessing the initial-state space of a random experiment.

(N) Von Plato (1994: ch. 5) gives an historical overview of the method of arbitrary functions. A modern treatment of the mathematics is Engel (1992). Stevens (2003) is by far the most comprehensive contemporary investigation of the range approach. Remarkable about his treatment is that he sticks to a realistic modelling insofar as he avoids the transition to limiting cases that is typical for mathematically-oriented treatises. He sketches applications of the method to statistical physics (ch. 4.8) and population ecology (ch. 4.9) and investigates the presuppositions under which it could work in the social sciences (ch. 5.4). According to him, there is good reason to view probability as a complex high-level phenomenon that occurs if and only if initial-state spaces are structured in the indicated way. Consequently, he is inclined to reject the 'simple probabilities' of the single-case propensity theory (ch. 5.6).

(O) I have little to add to these comprehensive and thorough studies, except concerning the question of the interpretation of the emerging probabilities. While the early writers like Poincaré and Hopf adhered to a frequency theory of probability and accordingly understood the density functions as idealized representations of actual frequencies, Strevens deliberately remains neutral in view of interpretational questions (Strevens, 2003: ch. 1.32, 1.33, 2.14, 2.3). He is interested in the 'physics', not the 'metaphysics', of probability. According to him, you may interpret the density functions in any way you like, whereupon your interpretation will simply carry over to the resulting outcome probabilities. The situation is: probabilities in – probabilities out; the density functions over the initial-state space represent probability distributions. In general, one can say that the phenomenon caught by the phrase 'method of arbitrary functions' or 'range approach', if discussed at all by writers on probability, was taken to be an *explanation* for the existence of probabilistic patterns, but not as an *interpretation* of the concept of probability. In contrast to that, I would like to take seriously the idea that the range approach provides an independent proposal for an interpretation of probability statements, in the sense that they are made true or false by the structure of the corresponding initial-state spaces.

## 5.5 The main objection to the range interpretation

The range conception identifies probabilities with proportions of outcomes in suitably structured initial-state spaces. This is, furthermore, not just meant to be a contingent numerical identity, but a proposal for an objective interpretation of probability. What I am going to discuss now under the heading 'main objection to the range interpretation' has two parts. The first does not yet challenge the numerical identification, but just denies that it is suitable for an interpretation of probability statements. The second calls into doubt the numerical identification, which is, of course, a minimal requirement without which you cannot even start to consider the range proposal. The objection goes like this:

As with the classical approach, we do not get an interpretation of probability, because the concept is in fact presupposed. The range approach has to assume that initial states in the same small equilateral interval are approximately 'equally possible', i.e. it must assume an approximately uniform probability distribution over any such interval. Otherwise, there would be no reason to identify the proportion with which a certain outcome is represented within an arbitrary (but not too small) such interval with the probability of that outcome. The point is even more obvious with the method-of-arbitrary-functions formulation: What are those density functions supposed to be? Clearly, they represent probability distributions over initial states, and so the question of the interpretation of probability is merely shifted back, from probabilities of outcomes to probabilities of initial states.

Furthermore, it is not even possible to claim that the outcome probabilities are *numerically* identical to the respective proportions in the initial-state space. The probabilities of the outcomes depend not only on this space, but also on the actual density function over it. An initial-state space plus a probability density on it fix outcome probabilities, the space by itself fixes nothing. There are, of course, cases in which all ordinary density functions lead to roughly the same outcome probabilities, and in which only quite eccentric densities would yield different ones. But this does not mean that in these cases you can fix the probabilities without taking into account densities over the initial-state space, at least implicitly. For the actual density might well be an eccentric one, in which case you would get the probabilities wrong. Therefore, it is not even possible to numerically identify outcome probabilities with proportions in initial-state spaces. They are not always identical, and if they are, this is contingently so.

This criticism is undoubtedly strong. But it underestimates the fact that the range approach refers precisely to situations in which the initial-state space is structured in a way that guarantees independence of the outcome probabilities from any particular density function. Well, not quite – we have to confine ourselves to not-too-eccentric densities. But the independence is very far-reaching. We need not assume a uniform distribution over the initial-state space, i.e. treat the initial states as equally possible cases, or something like that. By considering only continuous density functions with appropriately bounded variation, we treat adjacent small regions of the initial-state space as approximately equally possible, but in fact this does not seem to be a severe restriction. The values of the probabilities we get out of the range approach would only be wrong if there were such a thing as the true probability distribution over the initial-state space and if this distribution were eccentric. In that case we would not dismiss the range approach, however, but rather conclude that we had overlooked some nomologically relevant factor, which means either that we had got the initial-state space wrong, or that this space is not primordial.

This observation is the key to answering the objection. The laws of Nature determine the result, given the initial conditions, but they leave open what those initial conditions are. As *they* do not care for the initial conditions, and *we* cannot control them sufficiently, it can only be by accident if on repeated trials of $E$ an eccentric distribution over $S$ emerges. This would not give us a reason to change our expectations concerning the possible outcomes, i.e. we would maintain our judgments concerning the objective probabilities. But *if*, on the contrary, we somehow convince ourselves that there is something behind the observed eccentric distribution, i.e. that it can be relied on for future predictions, we would conclude that there must be some nomological factor we have overlooked: the laws of Nature actually *do* care, contrary to what we thought. This reaction shows, I suppose, that we are quite ready to analyse probability according to the range interpretation. We are not prepared to accept 'just so' a falling-apart of objective probabilities and the respective proportions in the associated initial-state space. We would look for an explanation, which would then change our modelling of the experimental situation and thereby enable us to uphold the identification.

Furthermore, if we are able to maintain the *numerical* identity between objective probabilities and proportions of outcomes in initial-state spaces, we can analyse the concept of probability accordingly, without circularity, because in (RC) there is no talk about equally possible cases, nor about densities or distributions over initial-state spaces. The probabilities depend only on the space itself, after all. This discussion bears heavily on the distinction between laws of Nature and initial conditions, or, in the wording of von Kries, between the 'nomological' and the 'ontological' (von Kries, 1886: ch. 4.4). As in remark (H), above, I cannot do full justice to this topic here, but it seems to be clear that we tend to make the distinction in a way that supports the range conception.

Another reply to the main objection, put forward by Strevens (2003: ch. 2.53), considers perturbations of density functions. An eccentric density over the initial-state space would give us different probabilities, to be sure, but any perturbation of this density that is not very slight is liable to change the probabilities again. The probabilities that are due to an eccentric density are not resilient, their values depend on the fact that this very density, and no other, is the true one. In contrast to this, the probabilities given by the standard densities are insensitive to changes in the density: this is precisely the upshot of the method of arbitrary functions. Now, it is very doubtful if there ever is such a thing as 'the true density' on the initial-state space. Why should Nature, in its choice of initial states, follow some particular density function, after all (provided that we took all relevant laws of Nature into account in modelling the initial-state space)? Thus we can say either that the objective outcome probabilities are given by the regular densities, which all lead to the same probabilities, or else that there are no objective probabilities at all. This confirms the foregoing point that if we could somehow rely on a certain eccentric density to be the true one, this would have to be due to some neglected nomological factor that guarantees the stability of this density. Without such a stabilizing factor the probabilities the density gives rise to could not properly be called 'objective'.

These answers to the objection are not yet completely satisfying, as shown by a second, closely related criticism. In considering the proportions with which the various possible outcomes are represented in the initial-state space, we assume a standard measure on this space. To be sure, any continuous transformation of this measure with appropriately bounded variation would lead to the same probabilities, but why not choose a devious or even completely gerrymandered measure? Take again the throwing of a die. As we have seen, if we were able to construct the initial-state space and to survey it in detail, we would find it consisting of very small regions, a kind of patches, each patch uniformly leading to a particular result. Now, in view of these patches, we might choose a measure that blows up just those patches which lead to the outcome 'six', at the expense of all other patches. If one metricizes the initial-state space in such a non-standard way, there may well be a fixed proportion of 'six' in any (not-too-) small equilateral interval, but this proportion would now exceed the true probability of 'six'. That probability has not changed, of course. The random experiment is the same as before, we merely represented the initial-state space attached to it in an unusual way, i.e. chose a strange measure. On what grounds can such a representation be excluded?

The connection with the foregoing discussion is that an appropriately distorted initial-state space, with the 'six'-patches inflated and the others shrunk in relation to them, amounts to the same as the choice of the standard measure plus a density that is suitably periodic with a period of the size of the patches. Now, I said that *if* we should get the impression that Nature's choice of initial conditions could reliably be modelled by such a density function, we would conclude that we have overlooked some nomologically relevant factor and the initial-state space is not primordial. But could we not base our considerations on the suitably distorted space as well? And if we did, would we not then be forced to dismiss all *ordinary* density functions as eccentric, because any such function has troughs over the 'six'-patches of the distorted space? Would we not then, by an argument analogous to the foregoing one, be forced to conclude that we must have overlooked some nomologically relevant factor if we came to the conclusion that Nature followed *such* a density in selecting initial conditions? This would be a *reductio* of the reasoning given above. In this scenario, *any* density can appear to be regular or irregular, to be ordinary or eccentric, depending on the measure on the initial-state space. Therefore, if we rule out certain densities as irregular or eccentric, this can only be in relation to the standard measure, or to a large class of well-behaved transformations of the standard measure, but not to just any old measure one happens to choose. Consequently, it has to be assumed that standard ways of measuring the distance between vectors of physical quantities are distinguished in an objective sense. Nature has to single out the measure (or at least a large class of suitably well-behaved measures that are equivalent with respect to the relevant proportions) together with the initial-state space, if we are to get objective probabilities out of the range approach. It is then only in relation to this 'natural' measure (or class of measures) that a density counts as eccentric and thus as giving reason to remodel the physical situation by looking for a further, so far neglected nomological factor.

A distorting measure of the kind sketched above gives rise to a very unnatural representation of the possible initial states. The different components of the initial-state vector do not represent the usual physical magnitudes any longer, but related ones that are the result of complicated and entangled transformations of the former. (The components of the initial-state vector are not distorted one by one, but all together.) The new magnitudes we get in this way play no role in physics, and with good reason. For example, the usual conservation laws do not hold for them. Similar observations can be made for even more gerrymandered transformations of the initial-state space, if, e.g., the space is torn into pieces that would then be rearranged so as to yield different probabilities of the outcomes, or no probabilities at all.[6] The standard physical magnitudes would thereby be transformed in a not only complicated and entangled but even discontinuous way. In general, it should be possible to rule out as unnatural the magnitudes that are the result of such transformations, and with them the corresponding distance measures for initial states. This is sketchy and programmatic, of course. We cannot, however, be satisfied

---

[6] Strevens speaks of 'racked' and 'hacked' initial condition variables, respectively (Strevens 2003: ch. 2.53).

with just saying that if we represent the initial-state space in standard 'reasonable' ways, we always get the same proportions of the possible outcomes. The laws of Nature do not only have to tell us what the outcome of the experiment is, given certain initial conditions, but also which distance measure, or class of equivalent measures, to choose for initial states.[7] To put it the other way round: as far as the choice of the measure is a matter of convention, as far as the initial-state space has no 'naturally-built-in' measure, or class of measures that are equivalent regarding the proportions of the outcomes, the probabilities we get out of the range approach are merely conventional as well.

That such an assumption is necessary can also be seen with Strevens's perturbation argument. The problem here is that a certain regular measure for the space of possible densities over the initial-state space, i.e. a distance measure for densities, is tacitly assumed. Otherwise we could not say that *small* perturbations of a density are likely to change the probabilities if the density is eccentric, but not if it is ordinary. It should be possible in principle to select a measure for densities that is itself eccentric in such a way so as to render the foregoing statement wrong. Again, we have to assume 'natural' distance measures for densities to be able to assess the effects of 'small' perturbations of densities.

To resume: The main objection to the range conception can be answered if and only if standard ways of measuring the distance between initial-state vectors are naturally distinguished. The probabilities we get out of the range conception are objective precisely to the extent in which this is the case. But in view of the serious difficulties of propensity and frequency theories, the range approach seems to be a viable option for an objective interpretation of probability.[8]

## References

Eagle, A. (2004). Twenty-one arguments against propensity analyses of probability. *Erkenntnis*, **60**, 371–416.
Engel, E. (1992). *A Road to Randomness in Physical Systems*. Berlin: Springer-Verlag.
Guttmann, Y. (1999). *The Concept of Probability in Statistical Physics*. Cambridge: Cambridge University Press.

---

[7] See, however, Guttmann (1999: ch. 4.8), for a sceptical attitude concerning this program in the context of statistical mechanics.
[8] Note of acknowledgement: A shorter version of this chapter was presented at the Munich Conference on Time, Chance, and Reduction (8–10 March 2006). I am very grateful to Gerhard Ernst and Andreas Hüttemann for giving me the opportunity, and to the conference participants for helpful discussion. Much of the chapter was written about a year before during a stay at the University of Konstanz Center for Junior Research Fellows. Here, I am very grateful to Luc Bovens and Stephan Hartmann for inviting me, to the Alexander-von-Humboldt-Stiftung for providing a three-month scholarship, and to the members of the Philosophy, Probability, and Modeling Group for providing several instructive criticisms. Last but not least I am grateful to Christopher von Bülow for improving my English.

Hájek, A. (1997). 'Mises redux' – redux: Fifteen arguments against finite frequentism. *Erkenntnis*, **45**, 209–227.

Hájek, A. (2009). Fifteen arguments against hypothetical frequentism. *Erkenntnis*, **70**, 211–235.

Heidelberger, M. (2001). Origins of the logical theory of probability: von Kries, Wittgenstein, Waismann. *International Studies in the Philosophy of Science*, **15**, 177–188.

Hopf, E. (1934). On causality, statistics and probability. *Journal of Mathematics and Physics*, **13**, 51–102.

Hopf, E. (1936). Über die Bedeutung der willkürlichen Funktionen für die Wahrscheinlichkeitstheorie. *Jahresbericht der Deutschen Mathematiker-Vereinigung*, **46**, 179–195.

Poincaré, H. (1896). *Calcul des Probabilités*, Paris.

Poincaré, H. (1902). *La Science et l'Hypothèse*, Paris.

Strevens, M. (2003). *Bigger than Chaos*. Cambridge, MA: Harvard University Press.

Von Kries, J. (1886). *Die Principien der Wahrscheinlichkeitsrechnung*. Tübingen: Mohr Siebeck (2nd unchanged printing 1927).

Von Plato, J. (1994). *Creating Modern Probability*. Cambridge: Cambridge University Press.

Von Wright, G. H. (1982). Wittgenstein on probability. In *Wittgenstein*, ed. G. H. von Wright. Oxford: Blackwell.

# 6
# Probability in Boltzmannian statistical mechanics

ROMAN FRIGG

## 6.1 Introduction

A cup of tea, left to its own, cools down while the surrounding air heats up until both have reached the same temperature, and a gas, confined to the left half of a room, uniformly spreads over the entire available space as soon as the confining wall is removed. Thermodynamics (TD) characterizes such processes in terms of an increase of thermodynamic entropy, which attains its maximum value at equilibrium, and the second law of thermodynamics posits that in an isolated system entropy cannot decrease. The aim of statistical mechanics (SM) is to explain the behaviour of these systems, in particular their conformity with the second law, in terms of the dynamical laws governing the individual molecules of which the systems are made up. In what follows these laws are assumed to be the ones of classical mechanics.

An influential suggestion of how this could be achieved was made by Ludwig Boltzmann (1877), and variants of it are currently regarded by many as the most promising option among the plethora of approaches to SM. Although these variants share a commitment to Boltzmann's basic ideas, they differ widely in how these ideas are implemented and used. These differences become most tangible when we look at how the different approaches deal with probabilities. There are two fundamentally different ways of introducing probabilities into SM, and even within these two groups there are important disparities as regards both technical and interpretational issues. The aim of this chapter is to give a statement of these approaches and point out wherein their difficulties lie.

## 6.2 Boltzmannian statistical mechanics

The state of a classical system of $n$ particles with three degrees of freedom is completely specified by a point $x = (\mathbf{r}_1, \mathbf{p}_1, \ldots, \mathbf{r}_n, \mathbf{p}_n)$ in its $6n$ dimensional phase space $\Gamma$, where $\mathbf{r}_i$ and $\mathbf{p}_i$ are position and momentum, respectively, of the $i$th particle.[1] For reasons that will become clear soon, $x$ is referred to as the system's (fine-grained) micro-state. $\Gamma$ is endowed with the 'standard' Lebesgue measure $\mu$. Because we only consider systems in which the total energy is conserved and which have finite spatial extension, only a finite region $\Gamma_a \subset \Gamma$ is accessible to $x$.

We now assume that dynamics of the system is governed by Hamilton's equations, which define a measure-preserving flow $\phi_t$ on $\Gamma$: for all times $t$, $\phi_t : \Gamma \to \Gamma$ is a one-to-one mapping and $\mu(R) = \mu(\phi_t(R))$ for all regions $R \subseteq \Gamma$.[2] To indicate that we consider the image of a point $x$ under the dynamics of the system between two instants of time $t_1$ and $t_2$ (where $t_1 < t_2$) I write '$\phi_{t_2-t_1}(x)$', and likewise for '$\phi_{t_2-t_1}(R)$'. The inverse is denoted by '$\phi_{t_1-t_2}(x)$' and provides the system's state at time $t_1$ if its state was $x$ at $t_2$, and likewise for '$\phi_{t_1-t_2}(R)$'.

It is one of the central assumptions of Boltzmannian SM (BSM) that the system's macro-states $M_i$, $i = 1, \ldots, m$ (and $m < \infty$) – characterized by the values of macroscopic variables such as pressure, volume, and temperature – supervene on the system's micro-states, meaning that a change in the macro-state must be accompanied by a change in the micro-state. Hence the system's micro-state uniquely determines the system's macro-state in that to every given $x \in \Gamma_a$ there corresponds *exactly one* macro-state, $M(x)$. This determination relation is not one-to-one as many different $x$ can correspond to the same macro-state. It is therefore natural to define the macro-region of $M_i$, $\Gamma_{M_i}$, as the subset of $\Gamma_a$ that consists of all micro-states corresponding to macro-state $M_i$:

$$\Gamma_{M_i} := \{x \in \Gamma_a \mid M_i = M(x)\}, i = 1, \ldots, m. \tag{6.1}$$

The $\Gamma_{M_i}$ together form a partition of $\Gamma_a$, meaning that they do not overlap and jointly cover $\Gamma_a$: $\Gamma_{M_i} \cap \Gamma_{M_j} = \varnothing$ for all $i \neq j$ and $i, j = 1, \ldots, m$, and $\Gamma_{M_1} \cup \ldots \cup \Gamma_{M_m} = \Gamma_a$, where '$\cup$', '$\cap$' and '$\varnothing$' denote set theoretic union, intersection and the empty set respectively.

The Boltzmann entropy of a macro-state $M$ is defined as

$$S_B(M) = k_B \log[\mu(\Gamma_M)], \tag{6.2}$$

where $M$ ranges over the $M_i$ and $k_B$ is the so-called Boltzmann constant. The macro state for which $S_B$ is maximal is the equilibrium state, meaning that the system is

---

[1] For an introduction to classical mechanics see, for instance, Goldstein (1981) and Abraham and Marsden (1980); Goldstein (2001), Goldstein and Lebowitz (2004) and Lebowitz (1993a, 1993b, 1999) provide compact statements of the Boltzmannian approach to SM.

[2] In a Hamiltonian system energy $E$ is conserved and hence the motion of the system is confined to the $6n-1$ dimensional energy hypersurface $\Gamma_E$ defined by the condition $H(x) = E$, where $H(x)$ is the Hamiltonian of the system. Restricting $\mu$ to $\Gamma_E$ yields a natural measure $\mu_E$ on this hypersurface. At some points in the argument below it would be more appropriate to use $\mu_E$ rather than $\mu$. However, in the specific circumstances this would need some explaining and as none of the conclusions I reach depend on this I will not belabour this point further. For further discussion of this point, see Frigg (2008: 103–113).

in equilibrium if it has reached this state.[3] For the sake of notational convenience we denote the equilibrium state by $M_{eq}$ and choose, without loss of generality, the labelling of the macro-states such that $M_m = M_{eq}$.

Given this, we can define the Boltzmann entropy of a *system* at time $t$ as the entropy of the system's macro-state at $t$:

$$S_B(t) := S_B(M_t) \qquad (6.3)$$

where $M_t$ is the system's macro-state at time $t$; i.e. $M_t := M(x(t))$, where $x(t)$ is the system's micro-state at $t$.

It is now common to accept the so-called past hypothesis, the postulate that the system starts off at $t_0$ in a low-entropy macro-condition, the 'past state' $M_p$ (and we choose our labelling of macro-states such that $M_1 = M_p$). How the past state is understood depends on one's views on the scope of SM. The grand majority of Boltzmannians take the system under investigation to be the entire universe and hence interpret the past state as the state of the universe just after the big bang (I will come back to this below). Those who stick to the spirit of laboratory physics regard laboratory systems as the relevant unit of analysis and see the past state as one that is brought about in some particular experimental situation (such as the gas being confined to the left half of the room). For the discussion in this chapter it is inconsequential which of these views is adopted.

The leading idea of BSM is that the behaviour of $S_B(t)$ should mirror the behaviour of the thermodynamic entropy $S_{TD}$; that is, it should increase with time $t$ and reach its maximum at equilibrium. We should not, however, expect that this mirroring be exact. The second law of theormodynamics is a universal law and says that $S_{TD}$ can never decrease. A statistical theory cannot explain such a law and we have to rest content if we explain the 'Boltzmannian version' of the second law (Callender, 1999), also referred to as 'Boltzmann's law' (BL):

Consider an arbitrary instant of time $t'$ and assume that at that time the Boltzmann entropy $S_B(t')$ of the system is low. It is then highly probable that at any time $t'' > t'$ we have $S_B(t'') \geq S_B(t')$.

A system that behaves in accordance with BL is said to exhibit 'thermodynamic like behaviour' (TD-like behaviour, for short).

What notion of probability is invoked in BL and what reasons do we have to believe that the claim it makes is true? Different approaches to BSM diverge in how they introduce probabilities into the theory and in how they explain the tendency

---

[3] This assumption is problematic for two reasons. First, Lavis (2005) has argued that associating equilibrium with *one* macro-state is problematic for different reasons. Second, that the equilibrium state is the macro-state with the highest Boltzmann entropy is true only for non-interacting systems like ideal gases (Uffink, 2007: section 4.4). I set these difficulties aside for now.

of $S_B(t)$ to increase. The most fundamental distinction is between approaches that assign probabilities directly to the system's macro-states, and approaches that assign probabilities to the system's micro-state being in a particular subset of the macro-region corresponding to the system's current macro-state; for want of better terms I refer to these as 'macro-probabilities' and 'micro-probabilities' respectively. I now present these approaches one at a time and examine whether, and if so to what extent, they succeed in explaining BL.

Before delving into the discussion, let me list those approaches to probability that can be discounted straight away within the context BSM, irrespective of how exactly probabilities are introduced into the theory. The first three items on this list are the classical interpretation, the logical interpretation and the so-called no-theory theory. These have not been put forward as interpretations of Boltzmannian probabilities, and this for good reasons. The first two simply are not the right kind of theories, while the no-theory does not seem to offer a substantial alternative to either the propensity theory or David Lewis' view (Frigg and Hoefer, 2007: 385, 386). Frequentism, as von Mises himself pointed out, is problematic as an interpretation of SM probabilities because a sequence of results that is produced by the same system does not satisfy the independence requirements of a collective (van Lith, 2001: 587). Finally, a propensity interpretation is ruled out by the fact that the underlying micro-theory, classical mechanics, is deterministic,[4] which is incompatible with there being propensities (Clark, 2001).

## 6.3 The macro-probability approach

In this section I discuss Boltzmann's (1877) proposal to assign probabilities to the system's macro-states and Paul and Tatiana Ehrenfest's (1912) view that these should be interpreted as time averages, based on the assumption that the system is ergodic.

### 6.3.1 Introducing macro-probabilities

Boltzmann's way of introducing probabilities into SM is intimately related to his construction of the macro-regions $\Gamma_{M_i}$, which is now also known as the 'combinatorial argument'.[5] Assume now that the system under investigation is a gas that consists of $n$ molecules of the same type and is confined to a finite volume $V$. Then consider the six-dimensional phase space of *one* gas molecule, commonly referred

---

[4] This is true of the systems studied in BSM; there are failures of determinism in classical mechanics when we look at a larger class of systems (Earman, 1986).
[5] For a presentation and discussion of this argument see Uffink (2007: 974–893).

to as $\mu$-space. The conservation of the total energy of the system and the fact that the gas is confined to $V$ results in only a finite part of the $\mu$-space being accessible to the particle's state. Now put a grid-like partition on $\mu$-space whose cells all have the size $\delta\omega$ and whose borders run in the directions of the momentum and position axes. This results in the accessible region being partitioned into a finite number of cells $\omega_i$, $i = 1, \ldots, k$. The state of the entire gas can be represented by $n$ points in $\mu$-space, every one of which comes to lie within a particular cell $\omega_i$. An 'arrangement' is a specification of which point lies in which cell; i.e. it is a list indicating, say, that the state of molecule 1 lies in $\omega_9$, the state of molecule 2 lies in $\omega_{13}$, and so on. A 'distribution' is a specification of how many points (no matter which ones) are in each cell; i.e. it is a $k$-tuple $(n_1, \ldots, n_k)$, expressing the fact that $n_1$ points are in cell $\omega_1$, and so on. Elementary combinatorial considerations show that each distribution is compatible with

$$W(n_1, \ldots, n_k) := \frac{n!}{n_1! \ldots n_k!} \tag{6.4}$$

arrangements. Boltzmann then regards the probability $p(n_1, \ldots, n_k)$ of a distribution as determined by $W(n_1, \ldots, n_k)$:

The probability of this distribution is then given by the number of permutations of which the elements of this distribution are capable, that is by the number $[W(n_1, \ldots, n_k)]$. As the most probable distribution, i.e. as the one corresponding to thermal equilibrium, we again regard that distribution for which this expression is maximal [...].

(Boltzmann, 1877: 187)[6]

In other words, Boltzmann's posit is that $p(n_1, \ldots, n_k)$ is proportional to $W(n_1, \ldots, n_k)$.

Macro-states are determined by distributions because the values of a system's macroscopic variables are fixed by the distribution. That is, each distribution corresponds to a macro-state.[7] To simplify notation, let us now assume that all distributions are labelled with an index $i$ and that there are $m$ of them. Then there corresponds a macro-state $M_i$ to every distribution $D_i$, $i = 1, \ldots, m$. This is sensible also from a formal point of view because each distribution corresponds to a particular region of $\Gamma$; regions corresponding to different distribution do not overlap and all regions together cover the accessible parts of $\Gamma$ (as we expect it to be

---

[6] This and all subsequent quotes from Boltzmann are my own translations. Square brackets indicate that Boltzmann's notation has been replaced by the one used in this chapter. To be precise, in the passage here quoted '$W$' refers to distribution over a partitioning of energy, not phase space. However, when Boltzmann discusses the partitioning of $\mu$-space a few pages further down (on p. 191) he refers the reader back to his earlier discussions that occur on p. 187 (quoted) and p. 176. Hence he endorses the posits made in this quote also for a partitioning of phase space.

[7] It may turn out that it is advantageous in certain situations to regard two or more distributions as belonging to the same macro-state (e.g. if the relevant macro-variables turn out to have the same values for all of them). As this would not alter any of the considerations to follow, I disregard this possibility henceforth.

the case for macro-regions, see Section 6.2). One can then show that the measure of each macro-region thus determined is given by

$$\mu(\Gamma_{M_i}) = W(n_1, \ldots, n_k)(\delta\omega)^n \qquad (6.5)$$

where $D_i = (n_1, \ldots, n_k)$ is the distribution corresponding to $M_i$.

This allows us to restate Boltzmann's postulate about probabilities in terms of the measures of the macro-regions $\Gamma_{M_i}$:

The probability of macro-state $M_i$ is given by

$$p(M_i) = c\,\mu(\Gamma_{M_i}), \; i = 1, \ldots, m \qquad (6.6)$$

where $c$ is a normalization constant determined by the condition $\sum_{i=1}^{m} p(M_i) = 1$.

I refer to this postulate as the 'proportionality postulate' and to probabilities thus defined as 'macro-probabilities'. The choice of the former label is evident; the latter is motivated by the fact that the postulate assigns probabilities to the macro-states of the system and that the values of these probabilities are determined by the measure of the corresponding macro-region. It is a consequence of this postulate that the equilibrium state is the most likely state.

Boltzmann does not provide an interpretation of macro-probabilities and we will return to this problem below. Let us first introduce his explanation of TD-like behaviour. The leading idea of Boltzmann's account of non-equilibrium SM is to explain the approach to equilibrium by a tendency of the system to evolve from an unlikely macro-state to a more likely macro-state and finally to the most likely macro-state:

In most cases the initial state will be a very unlikely state. From this state the system will steadily evolve towards more likely states until it has finally reached the most likely state, i.e. the state of thermal equilibrium.

*(Boltzmann, 1877: 165)*

[...] the system of bodies always evolves from a more unlikely to a more likely state.

*(Boltzmann, 1877: 166)*

In brief, Boltzmann's answer to the question of how SM explains the second law is that it lies in the nature of a system to move from states of lower towards states of higher probability.

### 6.3.2 Macro-probabilities scrutinized

The question now is where this tendency to move towards more probable states comes from. It does not follow from the probabilities themselves. The $p(M_i)$ as defined by Equation (6.6) are unconditional probabilities, and as such they do not imply anything about the succession of macro-states, let alone that ones of low probability are followed by ones of higher probability. As an example consider a loaded die; the probability to get a 'six' is 0.25 and all other numbers of spots have probability 0.15. Can we then infer that after, say, a 'three' we have to get a 'six' because the six is the most likely event? Of course not; in fact, we are much more likely not to get a 'six'. And the situation does not change if, to make the scenario more like SM, the dice is so strongly biased that getting a 'six' is by far the most likely event. If the probability for this to happen is 0.999, say, then we are of course very likely to see a 'six' when throwing the die next, but this has nothing to do with the fact that we threw a 'three' before. The expectation that we move through progressively more probable events on many rolls is simply a delusion.

A further (yet related) problem is that BL makes a statement about a *conditional* probability, namely the probability of the system's macro-state at $t''$, $M(t'')$, being such that $S_B(t'') > S_B(t')$, *given* that the system's macro-state at an earlier time $t'$ was such that its Boltzmann entropy was $S_B(t')$. The probabilities of the proportionality postulate are not of this kind. But maybe this mismatch is only apparent because conditional probabilities can be obtained from unconditional ones by using $p(B|A) = p(B\&A)/p(A)$, where $A$ and $B$ are arbitrary events and $p(A) > 0$. The relevant probabilities then would be $p(M_j|M_i) = p(M_j \& M_i)/p(M_i)$. Things are not that easy, unfortunately. $M_j$ and $M_i$ are mutually exclusive (the system can only be either in $M_j$ or $M_i$ but not in both) and hence the numerator of this relation is always zero.

The problem with this attempt is, of course, that it does not take time into account. What we really would have to calculate are probabilities of the form $p(M_j$ at $t''$ & $M_i$ at $t')/p(M_i$ at $t')$. The problem is that it is not clear how to do this on the basis of the proportionality postulate as time does not appear in this postulate at all. One way around this problem might be to slightly revise the postulate by building time dependence into it: $p(M_i$ at $t) = c\mu(\Gamma_{M_i}), i = 1, \ldots, m$. But even if this was justifiable, it would not fit the bill because it remains silent about how to calculate $p(M_j$ at $t''$ & $M_i$ at $t')$. In sum, the probabilities provided to us by the proportionality postulate are of no avail in explaining BL.

This becomes also intuitively clear once we turn to the problem of interpreting macro-probabilities. Boltzmann himself remains silent about this question and the interpretation that has become the standard view goes back to the Ehrenfests (1912). They suggest to interpret the $p(M_i)$ as time averages, i.e. as the fraction of time that the system spends in $\Gamma_{M_i}$.[8] The equilibrium state, then, is the most

---

[8] Although time averages are, loosely speaking, the 'continuum version' of frequencies, there are considerable differences between the two interpretations. For a discussion of this point see von Plato (1988: 262–65; 1989: 434–37).

probable state in the sense that the system is in equilibrium most of the time. Two things are worth noticing about this interpretation. First, it makes it intuitively clear why macro-probabilities cannot explain TD-like behaviour: from the fact that the system spends a certain fraction of time in some macro-state $M_i$ nothing follows about which state the system assumes after leaving $M_i$ – time averages have no implications for the succession of states. Second, this interpretation needs to be justified by a particular dynamical explanation; whether the time the system spends in a macro-state is proportional to the volume of the corresponding macro-region, depends on the dynamics of the system.

Of what kind does that dynamics have to be for a time average interpretation to be possible? Again, Boltzmann himself remains silent about the dynamical conditions necessary to backup his take on probability.[9] The Ehrenfests (1912) fill this gap by attributing to him the view that the system has to be ergodic. Roughly speaking, a system is ergodic on $\Gamma_a$ if for almost all trajectories, the fraction of time a trajectory spends in a region $R \subseteq \Gamma_a$ equals the fraction of the area of $\Gamma_a$ that is occupied by $R$.[10] If the system is ergodic, it follows that the time that its actual micro-state spends in each $\Gamma_{M_i}$ is proportional to $\mu(\Gamma_{M_i})$, which is what we need.

This proposal suffers from various technical difficulties having to do with the mathematical facts of ergodic theory. These problems are now widely known and need not be repeated here.[11] What needs to be noted here is that even if all these problems could be overcome, we still would not have explained BL because macro-probabilities, no matter how we interpret them, are simply the wrong probabilities to explain TD-like behaviour.[12]

Or would we? When justifying the time average interpretation we introduced the assumption that the system is ergodic. Ergodicity implies that a system that starts off in the past state sooner or later reaches equilibrium and stays in equilibrium most of the time. Is that not enough to rationalize BL? No it is not. What needs to be shown is not only that the system sooner or later reaches equilibrium, but also that this approach to equilibrium is such that whenever the system's macro-state changes the entropy is overwhelmingly likely to increase. To date, no one succeeded in showing that anything of that kind follows from ergodicity (or as Jos Uffink (2007: 981) puts the point, no proof of a statistical $H$-theorem on the basis of the ergodic hypothesis has been given yet). And there are doubts whether

---

[9] And, as Uffink (2007: 981) points out, he reaffirmed later in 1881 that he did not wish to commit to any dynamical condition in his 1877 paper.

[10] For a rigorous introduction to ergodic theory see Arnold and Avez (1968), and for an account of its long and tangled history Sklar (1993: Chs. 2 and 5) and von Plato (1994: Ch. 3).

[11] See Sklar (1993: Ch. 5), Earman and Rédei (1996) and van Lith (2001) for discussions.

[12] In line with the majority of Boltzmannians I here frame the problem of SM as explaining why entropy increases when a system is prepared in a low-entropy state. Lavis (2005: 254–61) criticizes this preoccupation with 'local' entropy increase as misplaced and suggests that what SM should aim to explain is that the Boltzmann entropy is close to its maximum most of the time.

such a proof is possible. Ergodicity is compatible with a wide range of different dynamical laws and it is in principle conceivable that there is a dynamical law that is such that the system passes through different macro-states in a way that is non-TD-like. To rule this out (and thereby justify BL on the basis of the ergodic hypothesis), one would have to show that the macro-state structure defined by the combinatorial argument is such that there exists no ergodic system that passes these macro-states in a non-TD-like way. It seems unlikely that there is an argument for this conclusion.

Even though ergodicity itself does not fit the bill, an important lesson can be learned from these considerations, namely that the key to understanding the approach to equilibrium lies in dynamical considerations. Assume for a minute that ergodicity was successful in justifying TD-like behaviour. Then, the explanation of TD-like behaviour would be entirely in terms of the dynamical properties of the system and the structure of the macro-regions; it would be an account of why, given ergodicity, the system visits the macro-states in the 'right' (i.e. TD-like) order. The proportionality principle and the probabilities it introduces would play no role in this; an explanation of the system's behaviour could be given without mentioning probabilities once. Of course, ergodicity is not the right condition, but the same would be the case for any dynamical condition. What does the explaining is an appeal to features of the system's phase flow in relation to the partitioning of the phase space into macro-regions $\Gamma_{M_i}$; probabilities have simply become an idle wheel.

## 6.4 The micro-probability approach

David Albert (2000) proposed a different approach to probabilities in BSM. In this section I discuss this approach and Barry Loewer's suggestion (2001; 2004) to understand these probabilities as Humean chances in David Lewis' (1986; 1994) sense.

### 6.4.1 Introducing micro-probabilities

We now assign probabilities not to the system's macro-states, but to measurable subsets of the macro-region corresponding to the system's macro-state at time $t$. For lack of a better term I refer to these probabilities as 'micro-probabilities'. An obvious way of introducing micro-probabilities is the so-called 'statistical postulate' (SP):

Let $M_t$ be the system's macro-state at time $t$. Then the probability at time $t$ that the system's micro-state lies in $A \subseteq \Gamma_{M_t}$ is $p_t(A) = \mu(A)/\mu(\Gamma_{M_t})$.

Given this, we can now calculate the probabilities occurring in BL. Let $\Gamma^{(+)}_{M_{t'}}$ be the subset $\Gamma_{M_{t'}}$ consisting of all micro-states $x$ that between $t'$ and $t''$ evolve, under the dynamics of the system, into macro-regions corresponding to macro-states of higher entropy: $\Gamma^{(+)}_{M_{t'}} := \{x \in M_{t'} \mid \phi_{t''-t'}(x) \in \Gamma_+\}$, where $\Gamma_+ := \bigcup_{M_i \in M_+} \Gamma_{M_i}$ and $M_+ := \{M_i \mid S_B(M_i) \geq S_B(M_{t'}), i = 1, \ldots, m\}$. The probability that $S_B(t'') \geq S_B(t')$ then is equal to $p_{t'}(\Gamma^{(+)}_{M_{t'}}) = \mu(\Gamma^{(+)}_{M_{t'}})/\mu(\Gamma_{M_{t'}})$.

For Boltzmann's law to be true the condition that $\mu(\Gamma^{(+)}_{M_t})/\mu(\Gamma_{M_t}) \geq 1 - \varepsilon$, where $\varepsilon$ is a small positive real number, must hold for all macro-states. Whether or not this is the case in a particular system is a substantial question, which depends on the system's dynamics. However, even if it is, there is a problem. It follows from the time reversal invariance of Hamilton's equations of motion that if it is true that the system is overwhelmingly likely to evolve towards a macro-state of higher entropy in the future, it is also overwhelmingly likely to have evolved into the current macro-state *from* a past macro-state $M''$ which also has *higher entropy*.[13]

Albert (2000: 71–96) discusses this problem at length and suggests fixing it by first taking the system under investigation to be the entire universe and then adopting the so-called past hypothesis (PH), the postulate that

[...] the world came into being in whatever particular low-entropy highly condensed big-bang sort of macrocondition it is that the normal inferential procedures of cosmology will eventually present to us.

*(Albert, 2000: 96)*

This postulate is not without problems (see Winsberg (2004b) and Earman (2006) for discussions), and those who accept it disagree about its status in the theory (see Callender's and Price's contributions to Hitchcock (2004), which present opposite views). I assume that these issues can be resolved in one way or another and presuppose PH in what follows. The problem with wrong retrodictions can then be solved, so Albert suggests, by conditionalizing on PH:

[...] the probability distribution that one ought to plug into the equations of motion in order to make inferences about the world is the one that's uniform, on the standard measure, over those regions of the phase space of the world which are compatible both with *whatever it is that we may happen to know about the present physical condition of the universe* (just as the *original* postulate [SP]) *and* with the hypothesis that the original macrocondition of the universe was the one associated with the *big bang*.

*(Albert, 2000: 95–96)*

---

[13] A point to this effect was first made by Ehrenfest and Ehrenfest (1907; 1912: 32–34). However, their argument is based on an explicitly probabilistic model and so its relevance to deterministic dynamical systems is tenuous.

From a technical point of view, this amounts to replacing SP with what I call the 'past hypothesis statistical postulate' (PHSP):

Let $M_t$ be the system's macro-state at time $t$. SP is valid for the past state $M_p$, which obtains at time $t_0$. For all times $t > t_0$ the probability at time $t$ that the system's micro-state lies in $A$ is

$$p_t(A|R_t) = \frac{\mu(A \cap R_t)}{\mu(R_t)} \tag{6.7}$$

where $R_t := M_t \cap \phi_{t-t_0}(M_p)$.

Again, whether or not BL is true given this postulate is a substantive question having to do both with the construction of the macro-states as well as the dynamics of the system; I will come back to this issue below. Let us first turn to the question of how these probabilities can be interpreted.

### 6.4.2 Humean chance

The basis for Lewis' theory of probability is the so-called Humean mosaic, the collection of all non-modal and non-probabilistic actual events making up the world's entire history (from the very beginning to the very end) upon which all other facts supervene. Lewis himself suggested that the mosaic consists of space-time points plus local field quantities representing material stuff. In a classical mechanical system the Humean mosaic simply consists of the trajectory of the system's micro-state in phase space, on which the system's macro-states supervene.

The next element of Lewis' theory is a thought experiment. To make this explicit – more explicit than it is in Lewis' own presentation – I introduce a fictitious creature, Lewis' Demon. In contrast to human beings who can only know a small part of the Humean mosaic, Lewis' Demon knows the entire mosaic. The demon now formulates various deductive systems which make true assertions about what is the case, and, perhaps, also about what the probability for certain events are. Then the demon is asked to choose the best among these systems. The laws of Nature are the true theorems of this system and the chances for certain events to occur are what the probabilistic laws of the best system say they are (Lewis, 1994: 480). Following Loewer (2004), I call probabilities thus defined *L*-chances.

The best system is the one that strikes the best balance between strength, simplicity and fit. The notions of strength and simplicity are given to the demon and are taken for granted in this context, but the notion of fit needs explicit definition. Every system assigns probabilities to certain courses of history, among them the actual course; the fit of the system is then measured by the probability that it assigns to the actual course of history, i.e. by how likely it regards the things to happen that actually do happen. By definition, systems that do not involve probabilistic laws have perfect fit. As an illustration, consider a Humean mosaic consisting only of ten outcomes of a coin flip: HHTHTTHHTT. Theory $T_1$ posits that all events are independent and sets $p(H) = p(T) = 0.5$; theory $T_2$ shares the independence assumption but posits $p(H) = 0.9$ and $p(T) = 0.1$. It follows that $T_1$ has better fit than $T_2$ because $(0.5)^{10} > (0.1)^5 (0.9)^5$.

Loewer's suggestion is that BSM as introduced above – the package of classical mechanics, PH and PHSP – is a putative best system of the sort just described and that PHSP probabilities can therefore be regarded as Humean chances:

Recall that (in Albert's formulation) the fundamental postulates of statistical mechanics are fundamental dynamical laws (Newtonian in the case of classical statistical mechanics), a postulate that the initial condition was one of low entropy, and the postulate that the probability distribution at the origin of the universe is the microcanonical distribution conditional on the low entropy condition. The idea then is that this package is a putative Best System of our world. The contingent generalisations it entails are laws and the chance statements it entails give the chances. It is simple and enormously informative. In particular, it entails probabilistic versions of all the principles of thermodynamics. That it qualifies as a best system for a world like ours is very plausible. By being part of the Best System the probability distribution earns its status as a law and is thus able to confer lawfulness on those generalisations that it (together with the dynamical laws) entails.

*(Loewer, 2001: 618; cf. 2004: 1124)*

There is an obvious problem with this view, namely that it assigns non-trivial probabilities (i.e. ones that can have values other than 0 and 1) to events within a deterministic framework, which Lewis himself thought was impossible (Lewis, 1986: 118). Loewer claims that Lewis was wrong on this and suggests that introducing probabilities via initial conditions solves the problem of reconciling determinism and chance:

[...] while there are chances different from 0 and 1 for possible initial conditions the chances of any event *A* after the initial time will be either 1 or 0 since *A*'s occurrence or non-occurrence will be entailed by the initial state and the deterministic laws. However, we can define a kind of dynamical chance which I call 'macroscopic chance'. The macroscopic chance at *t* of event *A* is the probability given by starting with the micro-canonical distribution over the initial conditions and then conditionalising on the entire macroscopic history of the world (including the low entropy postulate) up until *t*. [...] this probability distribution is completely compatible with deterministic laws since it concerns only the initial conditions of the universe.

*(Loewer, 2001: 618–19; cf. 2004: 1124)*

Loewer does not tell us what exactly he means by 'a kind of dynamical chance', in what sense this chance is macroscopic, how its values are calculated, and how it connects to the technical apparatus of SM. I will now present how I think this proposal is best understood and show that, on this reading, Loewer's 'macroscopic chances' coincide with PHSP as formulated above modulo a qualification to which I turn below.

As above, I take the system's state at $t > t_0$ to be the macro-state $M_t$. We now need to determine the probability of the event 'being in set $A \subseteq \Gamma_{M_t}$ at time $t$'. As I understand it, Loewer's proposal falls into two parts. The first is that the probability of an event at a time $t$ is 'completely determined' by the probability of the corresponding event at time $t_0$; that is, the probability of the event 'being in set $A$ at time $t$', $p_t(A)$, is equal to the probability of 'being in set $A_0$ at time $t_0$' where $A_0$ is, by definition, the set that evolves into $A$ under the dynamics of the system after time $t$ has elapsed. Formally, $p_t(A) = \mu_0(A_0) = \mu_0(\phi_{t_0-t}(A))$, where $\mu_0$ is the micro-canonical distribution over the past state, i.e. $\mu_0(\cdot) = \mu(\cdot \cap \Gamma_{M_p})/\mu(\Gamma_{M_p})$.

The second part is 'conditionalizing on the entire macroscopic history of the world [...] up to time $t$'. Loewer does not define what he means by 'macroscopic history', but it seems natural to take a macro-history to be a specification of the system's macro-state at *every* instant of time between $t_0$ and $t$. A possible macro-history, for instance, would be that the system is in macro-state $M_1$ during the interval $[t_0, t_1]$, in $M_5$ during $(t_1, t_2]$, in $M_7$ during $(t_2, t_3]$, etc., where $t_1 < t_2 < t_3 < \cdots$ are the instants of time at which the system changes from one macro-state into another. What we are now expected to calculate is the probability of 'being in set $A$ at time $t$' *given* the system's macro-history. Let $Q_t$ be the set of all micro-states in $\Gamma_{M_t}$ that are compatible with the entire past history of the system; i.e. it is the set of all $x \in \Gamma_{M_t}$ that lie on trajectories that for every $t$ were in the $\Gamma_{M_k}$ corresponding to the *actual* macro-state of the system at $t$. The sought-after conditional probability then is $p_t(A \mid Q_t) = p_t(A \& Q_t)/p_t(Q_t)$, provided that $p_t(Q_t) \neq 0$, which, as we shall see, is the problematic condition.

Putting these two parts together we obtain the fundamental rule introducing L-chances for deterministic systems, which I call the L-chance statistical postulate (LSP):

Let $M_t$ be the system's macro-state at time $t$. SP is valid for the past state $M_p$, which obtains at $t_0$. For all times $t > t_0$ the probability that the system's micro-state lies in A is

$$p_t(A \mid Q_t) = \frac{\mu_0(\phi_{t_0-t}(A \cap Q_t))}{\mu_0(\phi_{t_0-t}(Q_t))} \quad (6.8)$$

where $Q_t$ is the subset of $M_t$ of micro-states compatible with the entire past history of the system and, again, $\mu_0(\cdot) = \mu(\cdot \cap \Gamma_{M_p})/\mu(\Gamma_{M_p})$.

The crucial thing to realize now is that due to the conservation of the Liouville measure the expression for the conditional probability in PHSP can be expressed as $p_t(A \mid R_t) = \mu(\phi_{t_0-t}(A \cap R_t))/\mu(\phi_{t_0-t}(R_t))$. Trivially, we can substitute $\mu_0$ for $\mu$ in this expression which makes it formally equivalent to Equation (6.8). Hence

PHSP can be interpreted as attributing probabilities to events at $t > t_0$ solely on the basis of the micro-canonical measure over the initial conditions, which is what Loewer needs.

However, the equivalence of PHSP and LSP is only formal because there is an important difference between $Q_t$ and $R_t$: $R_t$ only involves a conditionalization on PH, while $Q_t$ contains the *entire* past history. This difference will be important later on.

### 6.4.3 Problems with fit

Loewer claims that BSM as introduced above is the system that strikes the best balance between simplicity, strength and fit. Trivially, this implies that BSM can be ranked along these three dimensions. Simplicity and strength are no more problematic in SM than they are in any other context and I shall therefore not discuss them further here. The problematic concept is fit.

The fit of a theory is measured in terms of the probability it assigns to the actual course of history. But what history? Given that $L$-chances are calculated using the Lebesgue measure, which assigns measure zero to *any* trajectory, they do not lead to a non-trivial ranking of micro-histories (trajectories in $\Gamma$). The right choice seems to be to judge the fit of theory with respect to the system's macro-history.

What is the probability of a macro-history? A first answer to this question would be to simply use Equation (6.8) to calculate the probability of a macro-state at each instant of time and then multiply them all, just as we did in the above example with the coins, with the only difference that the probabilities are now not independent any more, which is accounted for in Equation (6.8). Of course, this is plain nonsense. There is an uncountable infinity of such probabilities and multiplying an uncountable infinity of numbers is an ill-defined operation. Determining the probability of a history by multiplying probabilities for the individual events in the history works fine as long as the events are discrete (like coin flips) but it fails when we have a continuum.

Maybe this was too crude a stab at the problem and when taking the right sorts of limits things work out fine. Let us discretize time by dividing the real axis into small intervals of length $\delta$, then calculate the probabilities at the instants $t_0$, $t_0 + \delta$, $t_0 + 2\delta$ etc., multiply them (there are only countably many now), and then take the limit $\delta \to 0$. This would work if the $p_t(A \mid Q_t)$ depended in a way on $\delta$ that would assure that the limit exists. This is not the case. And, surprisingly, the problem does not lie with the limit. It turns out that for all $t > t_1$ (i.e. after the first change of macro-state), the $p_t(A \mid Q_t)$ do not exist because $Q_t$ has measure zero, and this irrespective of $\delta$. This can be seen as follows. Take the above example of a macro-history and consider an instant $t \in (t_1, t_2]$ when the system is in macro-state $M_5$.

To calculate the probability of the system being in $M_5$ at $t$ we need to determine $Q_t$, the set of all micro-states in $\Gamma_{M_5}$ compatible with the past macro history. Now, these points must be such that they were in $M_1$ at $t_1$ and in $M_5$ just an instant later (i.e. for any $\varepsilon > 0$, at $t_1 + \varepsilon$ the system's state is in $\Gamma_{M_5}$). The mechanical systems we are considering have global solutions (or at least solutions for the entire interval $[t_0, t_f]$, where $t_f$ is the time when the system ceases to exist) and trajectories in such systems have finite phase velocity; that is, a phase point $x$ in $\Gamma$ cannot cross a finite distance in no time. From this it follows that the only points that satisfy the condition of being in $M_1$ at $t_1$ and in $M_5$ just an instant later are the ones that lie exactly on the boundary between $M_1$ and $M_5$. But the boundary of a $6n$ dimensional region is $6n - 1$ dimensional and has measure zero. Therefore $Q_t$ has measure zero for all $t > t_1$, and accordingly $p_t(A \mid Q_t)$ does not exist for $t > t_1$, no matter what $A$. Needless to say, this renders the limit $\delta \to 0$ obsolete.

The source of these difficulties seems to be Loewer's requirement that we conditionalize on the entire macro-history of the system. This suggests that the problem could be solved simply by reverting back to Albert's algorithm, which involves only conditionalizing on PH. Unfortunately the hope that this would make the problems go away is in vain. First, Loewer seems to be right that we need to conditionalize on the entire macro-history when calculating the fit of a system. Fit is defined as the probability that the system attributes to the entire course of actual history and hence the probability of an event at time $t$ must take the past into account. It can turn out to be the case that the evolution of a system is such that the probability of an event at $t$ is independent of (large parts of the) past of the system, in which case we need not take the past into account. But if certain conditional probabilities do not exist given the system's past, we cannot simply 'solve' the problem by ignoring it.

Second, even if one could somehow convincingly argue that conditionalizing on the entire past is indeed the wrong thing to do, this would not be the end of the difficulties because another technical problem arises, this time having to do with the limit. As above, let us discretize time by dividing the real axis into small intervals of length $\delta$ and calculate the relevant probabilities at the instants $t_0$, $t_0 + \delta$, $t_0 + 2\delta$ etc. The relevant probabilities are of the form $p(t', t'') := p(M_j \text{ at } t'' \mid M_i \text{ at } t' \,\&\, M_p \text{ at } t_0)$, where $t'$ and $t''$ are two consecutive instances on the discrete time axis (i.e. $t'' > t'$ and $t'' - t' = \delta$), and $M_i$ and $M_j$ are the system's macro-states at $t'$ and $t''$ respectively. Calculating these probabilities using PHSP we find:

$$p(t', t'') = \frac{\mu[\phi_{t'-t''}(\Gamma_{M_j}) \cap \Gamma_{M_i} \cap \phi_{t'-t_0}(\Gamma_{M_p})]}{\mu[\Gamma_{M_i} \cap \phi_{t'-t_0}(\Gamma_{M_p})]} \tag{6.9}$$

These probabilities exist and we then obtain the probability of the actual course of history by plugging in the correct $M_i$'s and multiply all of them.

The next step is to take the limit $\delta \to 0$, and this is where things go wrong. We need to distinguish two cases. First, the system is in the same macro-state at $t'$ and $t''$ (i.e. $M_i = M_j$). In this case $\lim_{\delta \to 0} p(t', t'') = 1$ because $\lim_{\delta \to 0} \phi_{t'-t''}(\Gamma_{M_i}) = \Gamma_{M_i}$. Second, the system is in two different macro-states at $t'$ and $t''$. In this case $\lim_{\delta \to 0} p(t', t'') = 0$ because $\lim_{\delta \to 0} \phi_{t'-t''}(\Gamma_{M_j}) = \Gamma_{M_j}$ and, by definition, $\Gamma_{M_j} \cap \Gamma_{M_i} = \emptyset$. Hence the product of all these probabilities always comes out zero in the limit, and this for any phase flow $\phi$ and for any macro-history that involves at least one change of macro-state (i.e. for any macro-history in which the system ever assumes a state other than the past state). Fit, calculated in this way, fails to put different systems into a non-trivial fit ordering and is therefore useless.

The moral to draw from this is that we should not take the limit and instead calculate fit with respect to a finite number of instants of time; that is, we should calculate fit with respect to a discrete macro-history (DMH) rather than a continuous one. By a DMH I mean a specification for a finite number of instants of time $t_0 =: \tau_0 \leq \tau_1 \leq \ldots \leq \tau_{j-1} \leq \tau_j := t_f$ (which can but need not be equidistant as the ones we considered above) of the system's macro-state at these instants: $M_p$ at $\tau_0$, $M_{\tau_1}$ at $\tau_1$, $M_{\tau_2}$ at $\tau_2$, ..., $M_{\tau_{j-1}}$ at $\tau_{j-1}$ and $M_{\tau_j}$ at $\tau_j$, where $M_{\tau_i}$ is the system's macro-state at time $\tau_i$.[14]

Before discussing the consequences of this move further, let me point out that once we go for this option, the seeming advantage of conditionalizing only on PH rather than the entire past history evaporates. If we only consider the system's macro-states at discrete times, $Q_t$ no longer needs to have measure zero. As a result, the probabilities introduced in Equation (6.8) are well defined. So, conditionalizing on the entire past is no problem as long as the relevant history is discrete.

How could we justify the use of a discrete rather than a continuous macro-history? We are forced into conditionalizing over a continuum of events by the conjunction of three assumptions: (1) the posit that time is continuous, (2) the assumption that the transition from one macro-state to another one takes place at a precise instant, (3) the posit that fit has to be determined with respect to the *entire* macro-history of the system. We have to give up at least one of these to justify the use of a discrete rather than a continuous macro-history. The problem is that all

---

[14] I assume $j$ to be finite. There is a further problem with infinite sequences (Elga, 2004). The difficulties I discuss in this section and the next are independent of that problem.

three elements either seem reasonable or are deeply entrenched in the theory and cannot be renounced without far-reaching consequences.

The first option, discretizing time, would solve the problem because if we assume that time is discrete the macro-history is discrete too. The problem with this suggestion is that it is *ad hoc* (because time in classical mechanics *is* continuous), and therefore defeats the purpose of SM. If we believe that classical mechanics is the fundamental theory governing the micro-constituents of the universe and set out to explain the behaviour of the universe in terms of its laws, not much seems to be gained if such an explanation can only be had at the expense of profoundly modifying these laws.

The second suggestion would be to allow for finite transition times between macro-states; that is, allowing for there to be periods of time during which it is indeterminate in which macro-state the system is. This suggestion is not without merit as one could argue that sharp transitions between macro-states are indeed a mathematical idealization that is ultimately unjustifiable from a physics perspective. However, sharp transitions are a direct consequence of the postulate that macro-states supervene on micro-states. This postulate is central to the Boltzmannian approach and it is not clear how it could be given up without unsettling the entire edifice of BSM.

The third option denies that we should conditionalize on the *complete* macro-history. The idea is that even though time at bottom is continuous, the macro-history takes record of the system's macro-state only at discrete instants and is oblivious about what happens between these. This is at once the most feasible and the most problematic suggestion. It is feasible because it does not require revisions in the structure of the theory. It is problematic because we have given up the notion that the fit of a theory has to be best with respect to the complete history of the world, and replaced it with the weaker requirement that fit be best for a partial history.[15] From the point of view of Lewis' theory this seems unmotivated. Fit, like truth, is a semantic concept characterizing the relation between the theory and the world, and if the Humean mosaic has continuous events in it there should still be a matter of fact about what the fit of the theory is.

Moreover, even if one is willing to believe that a discrete version of fit is satisfactory, it is not clear whether this leads to useful results. Depending on which particular instants of time one chooses to measure fit, one could get widely different results. These would be useful only if it was the case that the fit rankings came out

---

[15] And mind you, the point is not that the fit of the full history is in practice too complicated to calculate and we therefore settle for a more tractable notion; the point is that the fit of a complete macro-history is simply not defined because the relevant conditional probabilities do not exist.

the same no matter the choice of instants. There is at least a question whether this is the case.

### 6.4.4 The putative best system is not the best system

I now assume, for the sake of argument, that the use of a DMH to calculate fit can be defended in one way or another. Then a further problem emerges: the package consisting of Hamiltonian mechanics, PH and PHSP in fact is *not* the best system. The reason for this is that we can always improve the fit of a system if we choose a measure that, rather than being uniform over $\Gamma_{M_p}$, is somehow peaked over those initial conditions that are compatible with the entire DMH.

Let us make this more precise. As above, I first use Loewer's scheme and then discuss how conditionalizing à la Albert would change matters. The probability of the DMH is $p(DMH) = p_{\tau_0}(A_{\tau_0} | Q_{\tau_0}) \ldots p_{\tau_{j-1}}(A_{\tau_{j-1}} | Q_{\tau_{j-1}})$, where the $A_{\tau_i}$ are those subsets of $M_{\tau_i}$ that evolve into $M_{\tau_{i+1}}$ under the evolution of the system between $\tau_i$ and $\tau_{i+1}$. One can then prove that

$$p(DMH) = \mu_0[\Gamma_p \cap \phi_{\tau_0 - \tau_1}(\Gamma_1) \cap \ldots \cap \phi_{\tau_0 - \tau_j}(\Gamma_j)] \qquad (6.10)$$

where, for ease of notation, we set $\Gamma_p := \Gamma_{M_p}$ and $\Gamma_i := \Gamma_{M_{\tau_i}}$ for $i = 1, \ldots, j$. Now define $N := \Gamma_p \cap \phi_{\tau_0 - \tau_1}(\Gamma_1) \ldots \phi_{\tau_0 - \tau_j}(\Gamma_j)$. The fit of system is measured by the probability that it assigns to the actual DMH, which is given by Equation (6.10). It is a straightforward consequence of this equation that the fit of a system can be improved by replacing $\mu_0$ by a measure $\mu_P$ that is peaked over $N$, i.e. $\mu_P(N) > \mu_0(N)$ and $\mu_P(\Gamma_p \setminus N) < \mu_0(\Gamma_p \setminus N)$ while $\mu_P(\Gamma_p) = \mu_0(\Gamma_p)$. Fit becomes maximal (i.e. $p(DMH) = 1$) if, for instance, we choose the measure $\mu_N$ that assigns all the weight to $N$ and none to $\Gamma_p \setminus N$: $\mu_N(A) := k\mu_0(A \cap N)$, for all sets $A \subseteq \Gamma_p$ and $k = 1/\mu_0(N)$ (provided that $\mu_0(N) \neq 0$). Trivially, $N$ contains the *actual* initial condition of the universe. An even simpler and more convenient distribution that yields maximal fit is a Dirac delta function over the actual initial condition.

If there is such a simple way to improve (and even maximize) fit, why does the demon not provide us with a system comprising $\mu_N$ or a delta function? Coming up with such a system is not a problem for the demon, as, by assumption, he knows the entire Humean mosaic. One reason to prefer $\mu_0$ to other measures might be that these make the system less simple and that this loss in simplicity is not compensated by a corresponding gain in fit and strength. This seems implausible. Handling $\mu_N$ or a Dirac delta function rather than $\mu_0$ does not render the system much more complicated while the gain in fit is considerable. Hence simplicity does not seem to provide reason to prefer $\mu_0$ to other measures that have better fit.

This conclusion can be opposed on various grounds. The first is that the introduction of alternative distributions makes the system radically more complex and hence on balance the original system is better. Consider the delta function first, and assume for the sake of argument that $\Gamma_{M_p}$ is an interval. A counter to my claim then would be that the system with the delta function is complex because it has to store a number – the true initial condition – which is computationally costly because most numbers have decimal expansions that are infinitely long. By contrast, intervals can be stored cheaply; all the system has to store is something like 'the unit interval'.

This argument is unconvincing because it relies on a misleading choice of examples. A specification of an interval demands the specification of two numbers, the interval's boundaries. These *can*, of course, be natural numbers (and hence cheap to store), but this need not be the case; the boundaries of an interval can be real numbers. Moreover, it can also be the case that the true initial condition is a natural number, and hence cheap to store. So the bottom line is that storing an interval requires storing two numbers, while storing an initial condition only requires storing one number, and hence the claim that the system with the interval is simpler is unwarranted. Finally, this still leaves the question what happens if $\Gamma_{M_p}$ is not an interval. Things are less clear in that case, but arguably storing a complicated boundary is more costly than storing a single number.

What about a distribution peaked over $N$? The argument against this is that specifying $N$ is more complicated than specifying $\Gamma_{M_p}$ because the latter is given to us directly by the system's macro-state structure, while the former has to be determined on the basis of the entire Humean mosaic in an extra step. While this is true, it does not seem to be correct to say that this is an 'extra' in the sense that it involves carrying out a step that we would not otherwise have to carry out. As becomes clear from Equation (6.10), in determining the fit of the system we effectively determine $N$, and there is no way around determining the system's fit. So there is no extra work involved in determining $N$.

Another line of criticism claims that the gain in fit is not, as the above argument assumes, considerable.[16] In fact, so the argument goes, the original system has already extremely good fit and fiddling around with the measure would result at best in a small improvement of fit; and this small improvement is not enough to compensate the loss in simplicity which is involved in using distributions other than the original one. Therefore, the system containing $\mu_0$ is the best system after all.

I have already argued that I do not think that the use of alternative distributions makes the system more complex, but this criticism draws our attention to an

---

[16] Thanks to Eric Winsberg for a helpful discussion on this issue.

important point well worth discussing, namely the absence of a specification in the putative best system of the Hamiltonian of the universe.[17] Assume that the macro-history of the universe is indeed TD-like in that the entropy increases most of the time. Why then should we assume that the system consisting of Hamiltonian mechanics, PH and either PHSP or LSP has good fit? Whether or not this is the case depends on the dynamical properties of the Hamiltonian of the system. The phase flow $\phi$ – the solution of the equations of motion – explicitly occurs both in Equations (6.7) and (6.8) and hence whether these probabilities come out such that the system has good fit with the actual history depends on the properties of $\phi$. In fact, we cannot calculate any probabilities at all if we do not have $\phi$! Without a phase flow both PHSP and LSP remain silent and it is absolutely essential that we are told what $\phi$ to plug into the probabilistic algorithms if we want to obtain any results at all. So it is somewhat puzzling that Loewer does not say anything about $\phi$ at all and Albert (2000: 96) completely passes over it in his definitive statement of his view (he says something about it at another place; I come back to this below).

We are now faced with the situation that we have to plug some $\phi$ into either PHSP or LSP in order to make predictions, and at the same time we are not told anything about which $\phi$ to take. One way out of this predicament would be to assume that any $\phi$ would do. That is, the claim would be that every classical system with PH and either PHSP or LSP behaves in such a way that the actual history comes out having high fit. This, however, is wrong. Whether a system behaves in this way, and whether the system has good fit with the actual history of the world, essentially depends both on the dynamics of the system and the structure of the macro-regions. Some Hamiltonians do not give rise to an approach to equilibrium at all (e.g. a system of harmonic oscillators), and others may have constants of motion whose invariant surfaces prevent the system from moving into certain regions of phase space into which it would have to move for the macro-history to be what it is.

Hence, the best system cannot possibly consist only of the above three elements; it also has to contain a specification of the macro-state structure and a Hamiltonian (from which we obtain $\phi$ by solving the equations of motion). If we assume the macro-state structure to be given by the combinatorial argument, then the challenge is to present a Hamiltonian (or a class of Hamiltonians) for which the fit of the system is high. How would such a Hamiltonian look?

---

[17] Albert and Loewer regard Newtonian rather than Hamiltonian mechanics as part of the best system. Newtonian mechanics has a wider scope that Hamiltonian mechanics in that it allows also for dissipative force functions. However, such force functions do not occur in SM, where forces are assumed to be conservative. But the motions to which such forces give rise are more elegantly described by Hamilton mechanics. Hence I keep talking about Hamiltonian mechanics in what follows, which is not a drawback because the problems I describe would only be aggravated by considering full-fledged Newtonian mechanics.

Two criteria emerge from Albert's discussion. The first and obvious condition is that $\phi$ must be such that the probabilities calculated using PHSP come out right. The second, somewhat more involved, requirement is that $\phi$ be such that 'abnormal' micro-states (i.e. ones that lead to un-thermodynamic behaviour) are scattered in tiny clusters all over the macro-regions (Albert, 2000: 67, 81–85). The motivation for this condition is subtle and need not occupy us here; a detailed discussion can be found in Winsberg (2004a). Two things about these conditions are remarkable. First, Albert's discussion of these conditions suggests that they are technicalities that at some point have to be stated for completeness' sake and then can be put aside; in other words, his discussion suggests (although he does not say this explicitly) that these are simple conditions that hold true in all classical systems and once this fact is noticed we do not have to worry about them any more. This might explain why neither of them is mentioned in the final statement of his view (Albert, 2000: 96). This is misleading. These conditions are extremely special and many phase flows do not satisfy them. So it is very much an open question whether the Hamiltonians of those systems – in particular the universe as a whole – to which SM reasoning is applied satisfy these requirements.

To this one might reply that there is a sort of a transcendental argument for the conclusion that the Hamiltonian of the universe indeed must belong to this class of Hamiltonians, because if it did not then the universe simply would not be what it is and thermodynamics would not be true in it. The problem, and this is the second point, with this suggestion is that even if this argument was sound, it would not solve the original problem, namely that we cannot calculate any probabilities at all if we do not have $\phi$. The fact that the Hamiltonian of the universe satisfies all required criteria does not tell us what this Hamiltonian – or the phase flow $\phi$ associated with it – is, and as long as we do not have the Hamiltonian we simply cannot use either PHSP or LSP. Hence a system that does not contain the Hamiltonian of the universe is not really a system at all because the probabilistic algorithms require the phase flow $\phi$ of the universe to be given.

As it stands we do not know what $\phi$ is and hence the package of classical mechanics, PH and either PHSP or LSP fails to produce predictions at all. But assume now, again for the sake of argument, that we somehow manage to get hold of the Hamiltonian (and can also solve the equations of motion so that we obtain $\phi$). We could then say that the best system consists of Hamiltonian mechanics including the Hamiltonian of the universe, PH, the macro-state structure given by the combinatorial argument and a probabilistic algorithm. Before we can make this suggestion definitive we need to settle the question of which probabilistic algorithm to include. PHSP or LSP? There is again a trade-off between simplicity and fit. On the one hand, PHSP is simpler than LSP because conditionalizing on just the past state is simpler than conditionalizing on the entire discrete macro-history up to time $t$. On the other hand, LSP will generally yield better predictions than PHSP and hence make the system stronger. Whether on balance PHSP or LSP is the better choice depends, again, on $\phi$. It might be that the dynamics is so friendly that $R_t$ and $Q_t$ coincide, in which case PHSP is preferable on the grounds of simplicity. If this is not the case, then LSP may well be the better choice. Depending on which one comes out as striking the better balance, we add either PHSP or LSP to the above list. This then is our putative system.

Before raising a more fundamental objection against the view that the above quintet is a best system in the relevant sense, let me voice some scepticism about this line of defence against the original charge that fit can always be improved by substituting $\mu_0$ by a measure like $\mu_N$. The past state is defined *solely* via the value of some relevant macro-parameters and the corresponding macro-region, $\Gamma_{M_p}$, consists of *all* micro-states that are compatible with these values. In particular no facts about the behaviour of the $x \in \Gamma_{M_p}$ under the time evolution of the system is made. Now it would be something of a miracle if the overwhelming majority of points (relative to the Lebesgue measure) in a set *thus defined* would be the initial conditions of trajectories that are compatible with the *actual* course of history. We would have to have a very good and very special reason to assume that this is the case, and so far we do not have such a reason. And I doubt that there is one. For the approach to equilibrium to take place the system needs to be chaotic in some way, and such system involve divergence of nearby trajectories. For this reason it is plausible to assume that different $x \in \Gamma_{M_p}$ evolve into very different parts of phase space and thereby move through very different macro-regions, and hence give rise to different macro-histories. And for this reason using $\mu_N$ rather than $\mu_0$ would improve fit.

There is a more fundamental objection to the view that the quintet of Hamiltonian mechanics, PH, the combinatorial macro-state structure, a Hamiltonian satisfying some relevant condition and, say, LSP is a best system in Lewis' sense. The point is that a system that does not contain probabilities at all is much simpler than one that does, at least if probabilities are introduced either by PHSP or LSP. The laws are made by Lewis' demon and in the deterministic world of classical mechanics, the simplest system that the demon can come up with is one that consists of Hamilton's equations together with the Hamiltonian of the system, the macro-state structure, and the exact initial condition – everything else follows from this. In other words, simply adding the true initial condition as an axiom to mechanics and the macro-state structure is much simpler than adding probabilities via either PHSP or SPP – and doing this is no problem for the demon as he knows the exact initial condition.

Before arguing for this let me locate the moot point. It is important to be clear on the fact that the ability to solve equations is irrelevant to the argument, because both PHSP and LSP presuppose a solution to the equations of motion, because the phase flow $\phi$, which *is* the solution of the equations of motion, explicitly figures both in Equations (6.7) and (6.8).[18] In this respect systems comprising PHSP or LSP are not simpler than a system containing the exact initial condition as an axiom. So the controversy must be over the computational costs of handling initial conditions. The argument, then, must go something like this: having the laws and relevant macro-information is much simpler than knowing an initial condition, because storing an initial condition involves storing about $6n$ real numbers, which is costly because one has to take it in (presumably feed it into a computer), store it, and later on process it, which takes time and storage space and hence comes at a high cost of simplicity.

This is an illusion, which stems from the ambiguous use of 'macro-information'. Macro-information – knowledge about the values of macroscopic parameters such as pressure and volume – about the past state is useless when we want to apply, say, PHSP. What goes into this probabilistic algorithm is not macro-information *per se*; it is the set of *micro-states* compatible with the available macro-information. Once this is realized, the apparent advantage of a system containing probabilities evaporates because taking in and storing information about an entire set is more costly than taking in and storing a single point. This is because specifying a set in $6n$-dimensional space at least implies specifying its boundary, and this amounts to specifying a $(6n - 1)$-dimensional hypersurface, containing infinitely many points. And when applying PHSP we have to evolve forward in time this entire surface, which involves evolving forward in time infinitely many points. How could that possibly be simpler than taking in, storing, and evolving forward one single point?

In sum, there are strong prima facie reasons to assume that the system's fit can be improved by changing the initial measure. And even should it turn out that this is not the case, a system without probabilities would be better than one with probabilities. Therefore, neither PHSP nor LSP probabilities can be interpreted as Humean chances.

### 6.4.5 Interpreting BSM probabilities

Where does this leave us? I think that the above considerations make it plausible that what ultimately gives rise to the introduction of probabilities into classical mechanics are the epistemic limitations of those who use the theory. All we can know about a system is its macro-state and hence we put a probability distribution

---

[18] Notice that this is a very strong, and entirely unrealistic, assumption. Characteristically we do not have solutions to the equations of motion of systems studied in SM, and many physicists see this as the motivation to use statistical considerations to begin with (see for instance Khinchin, 1949).

over the micro-states compatible with that macro-state, which then reflects our lack of knowledge.

How these epistemic probabilities should be understood is a question that needs to be further discussed. Here I can only hint at two possibilities. The first, a version of objective Bayesianism, appeals to Jaynes' maximum entropy principle, which indeed instructs us to prefer $\mu_0$ to alternative measures because, given the information about the system's macro-state, $\mu_0$ maximizes the (continuous) Shannon entropy. The other alternative is to revise Lewis' account in a way that builds epistemic restrictions of the users of theories into systems. Hoefer's (2007) theory of Humean chance seems to make room for this possibility. This option should not scandalize the Humean. The bedrock of contemporary Humeanism as regards probability is the rejection of a metaphysical grounding of probabilities in propensities, tendencies, dispositions, and the like. At the same time the Humean regards probabilities as linked to human beliefs by the so-called Principal Principle, roughly the proposition that our subjective probabilities for an event to happen should match what we take to be its objective chance. Hence, for the Humean probabilities are both non-metaphysical and closely linked to beliefs. Hence, per se, the fact that SM probabilities have an epistemic component is no reason for outrage.

There are two main complaints about an epistemic interpretation of SM probabilities. The first[19] points out that the thermodynamic entropy is a property of a physical system and that $S_B$ coincides with it up to a constant. This, so the argument goes, is inexplicable on the basis of an epistemic approach to probabilities. This argument has indeed some force when put forward against the Gibbs entropy if the SM probabilities are given an ignorance interpretation, because the Gibbs entropy is defined in terms of the probability distribution of the system. However, it has no force against an epistemic interpretation of PHSP or LSP probabilities simply because these probabilities do not occur in the Boltzmann entropy, which is defined in terms of the measure of certain parts of phase space (see Equation (6.2)). Probabilities simply have nothing to do with it.

Another frequently heard objection is a complaint about the alleged causal efficacy of human knowledge. The point becomes clear in the following – rhetorical – questions by Albert:[20]

Can anybody seriously think that it is somehow *necessary*, that it is somehow *a priori*, that the particles that make up the material world must arrange themselves in accord with *what we know*, with what *we happen to have looked into*? Can anybody seriously think that our

---

[19] This point is often made in conversation, but I have been unable to locate it in print.
[20] Redhead (1995: 27–28, 32–33), Loewer (2001: 611), Goldstein (2001: 48), and Meacham (2005: 287–8) make essentially the same point.

merely being *ignorant* of the exact microconditions of thermodynamic systems plays some part in *bringing it about*, in *making it the case*, that (say) *milk dissolves in coffee*? How could that *be*?

(Albert, 2000: 64, original emphasis)[21]

It cannot be, and no one should think that it could. Proponents of epistemic probabilities need not believe in parapsychology, and therewith regard knowledge of tea and coffee as causally relevant to their cooling down.

What underlies this objection is the mistaken view that PHSP probabilities play a part in bringing about things in the world. Of course the cooling down of drinks and the boiling of kettles has nothing to do with what anybody thinks or knows about them; but they have nothing to do with the probabilities attached to these events either. Drinks cool down and kettles boil because the universe's initial condition is such that under the dynamics of the system it evolves into a state in which this happens. There is no causal connection between knowledge and happenings in the world, and, at least in the context of classical SM, nothing of that sort is suggested by an epistemic interpretation of probabilities.

We have now reached the same point as in the discussions of macro-probabilities in the previous section. All that is needed to explain why things happen is the initial condition and the dynamics. Prima facie appearances notwithstanding, neither PHSP nor LSP probabilities have any role to play in explaining why a system behaves as it does. If these probabilities come out as BL would have them, this is indicative, it is a symptom, of the system behaving 'thermodynamically', but it is not the cause for this behaviour.

## 6.5 Conclusion

I have discussed two different ways of introducing probabilities into BSM and argued that the first one is irredeemably flawed, while the second leads to probabilities that are best understood as having an epistemic component. The discussions of both approaches have shown that there is a 'blind spot' in the literature on BSM, namely dynamics. Too much emphasis has been placed on probabilities and not enough attention has been paid to the dynamical conditions that have to fall in place for a system to behave TD-like. This puts two items on the agenda of future discussions of BSM: we need to understand better the nature of the epistemic

---

[21] This contrasts starkly with the epistemic language that Albert uses throughout his discussion of PHSP. In the passages quoted above, for instance, he defines the past state as the one with which the 'normal inferential procedures of cosmology will eventually present us' and when discussing conditionalizing on PH he names 'make inferences about the world' as the aim and invites us to put a measure over those regions that are compatible with '*whatever* it is that we happen to know about the present physical condition of the universe'.

probabilities used in SM and we need to study more carefully the role that the dynamics of the system plays in explaining TD-like behaviour.[22]

## References

Abraham, R. and Marsden, J. E. (1980). *Foundations of Mechanics*, 2nd edn. London: Benjamin-Cummings.
Albert, D. (2000). *Time and Chance*. Cambridge, MA: Harvard University Press.
Arnold, V. and Avez, A. (1968). *Ergodic Problems in Classical Mechanics*. New York, Benjamin.
Boltzmann, L. (1877). Über die Beziehung zwischen dem zweiten Hauptsatze der mechanischen Wärmetheorie und der Wahrscheinlichkeitsrechnung resp. den Sätzen über das Wärmegleichgewicht. *Wiener Berichte*, **76**, 373–435. Reprinted in (1909). *Wissenschaftliche Abhandlungen*, vol. 2, ed. F. Hasenöhrl. Leipzig: J. A. Barth, pp. 164–223.
Callender, C. (1999) Reducing thermodynamics to statistical mechanics: the case of entropy. *Journal of Philosophy*, **96**, 348–373.
Clark, P. (2001). Statistical mechanics and the propensity interpretation of probability. In *Chance in Physics: Foundations and Perspectives*. ed. J. Bricmont *et al.* Berlin: Springer-Verlag, pp. 271–281.
Earman, J. (1986). *A Primer on Determinism*. Dordrecht: Kluwer.
Earman, J. (2006). The 'past hypothesis': not even false. *Studies in History and Philosophy of Modern Physics*, **37**, 399–430.
Earman, J. and Rédei, M. (1996). Why ergodic theory does not explain the success of equilibrium statistical mechanics. *British Journal for the Philosophy of Science*, **47**, 63–78.
Ehrenfest, P. and Ehrenfest, T. (1907). Über Zwei Bekannte Einwände gegen das Boltzmannsche *H*-Theorem. *Phyikalische Zeitschrift*, **8**, 311–14.
Ehrenfest, P. and Ehrenfest, T. (1912/1959). *The Conceptual Foundations of the Statistical Approach in Mechanics*. Mineola, NY: Dover (Reprinted 2002. First published in German in 1912; first English Translation 1959.)
Elga, A. (2004). Infinitesimal chances and the laws of nature. *Australasian Journal of Philosophy*, **82**, 67–76.
Frigg, R. and Hoefer, C. (2007). Probability in GRW theory. *Studies in the History and Philosophy of Modern Physics*, **38**, 371–389.
Frigg, R. (2008). A field guide to recent work on the foundations of statistical mechanics. In: *The Ashgate Companion to Contemporary Philosophy*. ed. D. Rickles. London: Ashgate, pp. 99–196.
Goldstein, H. (1981). *Classical Mechanics*. Reading, MA: Addison Wesley.
Goldstein, S. (2001). Boltzmann's approach to statistical mechanics. In *Chance in Physics: Foundations and Perspectives*, ed. J. Bricmont *et al.* Berlin: Springer-Verlag.
Goldstein, S. and Lebowitz, J. L. (2004). On the (Boltzmann) entropy of non-equilibrium systems. *Physica D: Nonlinear Phenomena*, **193** (1–4), 53–66.

---

[22] Thanks to Jossi Berkovitz, Nancy Cartwright, David Lavis, Carl Hoefer, Barry Loewer, Eric Winsberg, and Charlotte Werndl for helpful discussion and/or comments on earlier drafts. Thanks also to the audiences of the conference *Time, Chance, and Reduction* in Munich and the PSA meeting in Vancouver for feedback and suggestions.

Hitchcock, C. (ed.) (2004). *Contemporary Debates in Philosophy of Science*. Oxford: Blackwell.

Hoefer, C. (2007). The third way on objective probability: a skeptic's guide to objective chance. *Mind*, **116**, 549–596.

Khinchin, A. I. (1949). *Mathematical Foundations of Statistical Mechanics*. Mineola, NY: Dover.

Lavis, D. (2005). Boltzmann and Gibbs: an attempted reconciliation. *Studies in History and Philosophy of Modern Physics*, **36**, 245–273.

Lebowitz, J. L. (1993a). Boltzmann's entropy and time's arrow. *Physics Today*, September issue, 32–38.

Lebowitz, J. L. (1993b). Macroscopic laws, microscopic dynamics, time's arrow and Boltzmann's entropy. *Physica A*, **194**, 1–27.

Lebowitz, J. L. (1999). Statistical mechanics: a selective review of two central issues. *Reviews of Modern Physics*, **71**, 346–357.

Lewis, D. (1986). A Subjectivist's Guide to Objective Chance and Postscripts to 'A subjectivist's guide to objective Chance'. In *Philosophical Papers*, vol. 2, Oxford: Oxford University Press, pp. 83–132.

Lewis, D. (1994). Humean supervenience debugged. *Mind*, **103**, 473–90.

Loewer, B. (2001). Determinism and chance. *Studies in History and Philosophy of Modern Physics*, **32**, 609–629.

Loewer, B. (2004). David Lewis' Humean theory of objective chance. *Philosophy of Science*, **71**, 1115–1125.

Meacham, C. (2005). Three proposals regarding a theory of chance. *Philosophical Perspectives*, **19**, 281–307.

Redhead, M. (1995). *From Physics to Metaphysics*. Cambridge: Cambridge University Press.

Sklar, L. (1993). *Physics and Chance. Philosophical Issues in the Foundations of Statistical Mechanics*. Cambridge: Cambridge University Press.

van Lith, J. (2001). Ergodic theory, interpretations of probability and the foundations of statistical mechanics. *Studies in History and Philosophy of Modern Physics*, **32**, 581–594.

Uffink, J. (2007). Compendium of the foundations of classical statistical physics. In *Philosophy of Physics*, ed. J. Butterfield and J. Earman. Amsterdam: North-Holland, pp. 923–1047.

von Plato, J. (1988). Ergodic theory and the foundations of probability. In *Causation, Chance and Credence*, vol. 1, ed. B. Skyrms and W. L. Harper. Dordrecht: Kluwer, pp. 257–277.

von Plato, J. (1989). Probability in dynamical systems. In *Logic, Methodology and Philosophy of Science*, vol. VIII, ed. J. E. Fenstad, I. T. Frolov and R. Hilpinen. Amsterdam: North-Holland, pp. 427–443.

von Plato, J. (1994). *Creating Modern Probability*. Cambridge: Cambridge University Press.

Winsberg, E. (2004a). Can conditioning on the 'past hypothesis' militate against the reversibility objections? *Philosophy of Science*, **71**, 489–504.

Winsberg, E. (2004b). Laws and statistical mechanics. *Philosophy of Science*, **71**, 707–718.

# 7

# Humean metaphysics versus a metaphysics of powers

MICHAEL ESFELD

## 7.1 Humean metaphysics

Whereas the philosophy of science was dominated in the second half of the twentieth century by epistemological issues raised in the context of logical empiricism and its critics, the project of a metaphysics of Nature (metaphysics of science) has been rehabilitated recently. One can broadly distinguish three positions within that project: a Humean metaphysics that is close to empiricism (e.g. David Lewis, Barry Loewer and Helen Beebee), a metaphysics of universals (e.g. David Armstrong) and a metaphysics of powers (e.g. Sydney Shoemaker, Alexander Bird, Stephen Mumford as well as Charles Martin and John Heil). This chapter is about the opposition between the first and the third of these positions. I shall first outline the Humean view on properties, laws, causation and probabilities, and point out how on the one hand this view is parsimonious, whereas on the other hand it provokes the objection that it is deficient (this section). I shall explain how the metaphysics of powers seeks to remedy these deficiencies and consider its view of probabilities (Section 7.2). The chapter then recalls the standard argument against the conception of properties in Humean metaphysics, goes into arguments from physics and finally maintains that the metaphysics of powers, in contrast to Humean metaphysics, is able to do justice to both the ontological commitments of physics and of the special sciences (Section 7.3).

Humean metaphysics gets its name from the denial of the view that there are necessary connections between distinct entities in the world. Its most prominent formulation in contemporary philosophy is David Lewis' thesis of Humean supervenience:

It is the doctrine that all there is to the world is a vast mosaic of local matters of particular fact, just one little thing and then another. (...) We have geometry: a system of external relations of spatio-temporal distance between points. Maybe points of spacetime itself, maybe point-sized bits of matter or aether or fields, maybe both. And at those points we have local qualities: perfectly natural intrinsic properties which need nothing bigger than a point at which to be instantiated. For short: we have an arrangement of qualities. And that is all. There is no difference without difference in the arrangement of qualities. All else supervenes on that.

*(Lewis, 1986: ix–x)*

I shall take Lewis' statement to be characteristic of the position known today as Humean metaphysics. I shall not consider the relationship between that statement and the writings of David Hume. Humean metaphysics thus presupposes two things as primitive:

- the network of spatio-temporal relations between points of space-time. This is what unifies the world;
- the distribution of fundamental physical properties over the whole of space-time. Fundamental physical properties are all and only those physical properties that can occur at a point of space-time. Position, velocity, energy-mass, charge and spin count among the candidates for such properties. According to Lewis, these are natural, intrinsic and categorical properties.

Lewis' Humean metaphysics does not presuppose anything else as primitive. The description of everything else that there is in the world is to be derived from the description of the distribution of the intrinsic and categorical, fundamental physical properties over the whole of space-time. Everything else that there is in the world is a feature of that distribution.

Presupposing the distribution of the fundamental physical properties over the whole of space-time as primitive means that this distribution is entirely contingent. It is not only contingent that there is the distribution of fundamental physical properties as a whole that there is in fact, but each element in that distribution is also contingent. That is the point of denying necessary connections. Consequently, for each single token of a fundamental physical property at a space-time point, it is conceivable and metaphysically possible to hold that token (or its counterpart) fixed and to vary all the other tokens. Assuming that space-time and the distribution of fundamental physical properties in the Universe originate in the so-called big bang, given the big bang, in what manner the distribution of fundamental physical properties in the Universe develop is contingent. Given a possible world that is a duplicate of the big bang of the actual world, the development of the distribution of the fundamental physical properties in that possible world may be entirely different from the development of the distribution of the fundamental physical properties in the actual world. In short, the physical properties instantiated at any given space-time point or region do not impose any restrictions at all on the physical properties that can be instantiated at other space-time points or regions (Beebee, 2006).

By way of consequence, the fact that there are regularities in the distribution of the physical properties in space-time – and what these regularities are – has to be accepted as primitive. Certain of these regularities are *laws of Nature*. The most prominent proposal to distinguish the regularities that are laws within Humean metaphysics is the one of Ramsey–Lewis: the statements of laws of Nature are all and only the theorems of the system that achieves the best balance between simplicity and strength in empirical content in providing a complete description of the world (Lewis, 1973b: 72–75 and 1994: section 3). That view faces a number of objections, stemming notably from the concern that the criteria of simplicity and strength are subjective, being epistemic virtues, instead of objective, being anchored in the world. What is important for our purposes is that according to Humean metaphysics, there is first the distribution of fundamental physical properties over the whole of space-time, and then come the laws as supervening on that distribution. What the laws of Nature are, is hence fixed only at the end of the world.

The same goes for *causation*. Which property tokens stand in a relation of cause to effect supervenes on the distribution of the fundamental physical properties over the whole of space-time. According to a Humean regularity view of causation, all there is to causation are certain patterns of spatio-temporarily contiguous co-instantiations of properties over the whole of space-time. Hence, whether or not two given property tokens stand in a relation of cause to effect depends on what else there is in the world, namely on whether or not tokens of the same types as the two given ones are regularly co-instantiated.

Nothing of substance changes if one switches to a counterfactual account of causation such as the one proposed by David Lewis (1973a, 2004). Causation consists in certain counterfactual relations among property tokens. The statements describing these relations are made true by the distribution of the fundamental physical properties over the whole of space-time in the actual world (Loewer, 2007: 308–316). Lewis' theory of causation in terms of Humean supervenience does not commit one to accept his modal realism, that is, the view that other possible worlds exist, although finding out the truth-value of the counterfactual statements describing causal relations involves the comparison with other possible worlds. Whether or not a given counterfactual statement is true at the actual world depends on the laws holding at the actual world among other things. Hence, again, whether or not two given property tokens stand in a relation of cause to effect depends *only* on what else there is in the actual world, but it depends on *all* there is in the actual world, since the laws supervene on the distribution of the fundamental physical properties as a whole.

The fundamental physical properties are intrinsic and categorical. There are no *powers* or *dispositions* at space-time points over and above the intrinsic and categorical properties that occur at these points. Indeed, Humean metaphysics has no reason to admit dispositions at all. The standard Humean and empiricist account proceeds in two steps:

- a semantic step that consists in translating statements that ascribe dispositions to objects into counterfactual conditional statements;
- an ontological step consisting in the claim that these statements are made true by the distribution of the fundamental, intrinsic and categorical physical properties over the whole of space-time (and the laws of Nature – but the laws supervene on that distribution).

Consequently, given the second step, there is no need to recognize dispositions over and above categorical properties.

The conditional analysis of disposition ascriptions has come under attack in the nineties (Martin, 1994; Bird, 1998). That attack triggered a new debate on the ontological status of dispositions. But there are persuasive replies to that attack available that restore the conditional analysis of disposition ascriptions, starting with Lewis himself (Lewis, 1997 and then notably Gundersen, 2002; Cross, 2005: 323–325; Sparber, 2006; Choi, 2008). In any case, the conditional analysis of disposition ascriptions on its own is not able to decide the issue of the ontology of dispositions (Malzkorn, 2000). That issue depends on fundamental metaphysical considerations.

Nothing in Humean metaphysics hinders that the world can be *deterministic*. A Hume world can be such that, given complete knowledge of its initial conditions – e.g., the physical conditions at the big bang – and the laws of Nature, it is in principle possible to deduce the description of everything there is in the world. A Hume world thus is compatible with the letter of Laplacean determinism. However, the laws of Nature merely have a descriptive function. What the laws are is fixed only at the end of the world, since their supervenience basis is the distribution of the fundamental physical properties over the whole space-time. The statements of laws of Nature simply sum up at the end of the world what there is in the world.

One can therefore maintain that Humean metaphysics violates in any case the spirit of determinism, that is, the ideas that are associated with determinism: in a Hume world, the laws and the initial conditions do not determine the development of the distribution of physical properties in the sense that they are the ground for the existence of that development. Consequently, some Humeans hold that in a Hume world, there is no reason to raise concerns about whether or not determinism is compatible with free will: there is no such problem, since the laws depend on what there will be in the future and not the other way round (Beebee and Mele, 2002).

Humean metaphysics consequently faces the problem of induction in a metaphysical form: there is nothing in the past distribution of properties that could make true the prediction that the future distribution of properties will be like the past one. Lewis (1994: section 3) assumes that Nature is kind to us, that is, that there will be a few simple salient regularities that apply to the distribution of the physical properties throughout the whole Universe; but there is nothing in the past distribution of properties that could make such a claim about the whole Universe true. Of course, a Hume world can be such that the fundamental physical tokens in a limited region of space-time are de facto sufficient as a supervenience basis for the laws of Nature in the following sense: the best system describing that limited region of space-time is identical or comes close to the best system describing the whole world, since the distribution of the fundamental physical properties is uniform throughout the whole world. But the point is that there is nothing in the limited region of space-time considered – or in any region of space-time that is a proper part of the whole of space-time – that can make true the belief that space-time as a whole is like the proper part considered. Note that this is a point about the metaphysical relation of truth-making and not about the epistemological problem of the justification of beliefs acquired by induction (that latter problem remains even if – as in the metaphysics of powers – the former, metaphysical problem does not arise).

According to Humean metaphysics, the world can be deterministic in the explained sense, but it can also be probabilistic. In a Hume world, the regular patterns can be such that not all $F$s are followed by $G$s (supervenience basis for deterministic laws), but that only a certain proportion of the $F$s in the Universe are followed by $G$s (supervenience basis for probabilistic laws). If $F$ and $G$ are fundamental physical properties, these probabilities are irreducible and objective. The Humean account of probabilities is based on frequencies, but it is more sophisticated than a simple analysis in terms of relative frequencies, invoking the epistemic virtues of simplicity and strength that enter into the Humean best system account of the laws of Nature among others (Lewis, 1994; see Frigg and Hoefer, 2007 for an application to the quantum probabilities). However, for a Humean, the difference between a deterministic and a probabilistic world is not important as far as metaphysics is concerned, since every element in the distribution of the fundamental physical properties over the whole of space-time is contingent.

Consequently, the second law of thermodynamics (increase in entropy) simply describes a regularity pattern in the distribution of the fundamental physical properties in space-time. The same pattern can be described by statistical mechanics. If one takes for granted that there is a global increase in entropy, the description of the world by thermodynamics as well as by statistical mechanics presupposes that the state of the Universe immediately after the big bang is a state of extremely low entropy. This presupposition is known as the past hypothesis. The extremely low entropy at the beginning of the Universe is a contingent matter of fact. Nonetheless, on the Ramsey–Lewis view of the laws of Nature, it is possible to include the description of that fact among the theorems of the best system: admitting that description as theorem amounts to a considerable gain in simplicity (Callender, 2001: section 2.8). Furthermore, there is no intrinsic direction of time. The most widespread account within the framework of Humean metaphysics consists in saying that the direction of time supervenes on the distribution of fundamental physical properties which is such that there is an increase in entropy, and that increase in entropy accounts for the direction of time (see Loewer, 2007).

Hence, to sum up, on the one hand, Humean metaphysics is parsimonious because it accepts only the distribution of the fundamental physical properties as primitive and it conceives these properties as being intrinsic and categorical; on the other hand, there is in Humean metaphysics no answer to the question why the distribution of fundamental physical properties in the Universe developed as it did. Causation and laws supervene on that distribution; consequently, they cannot account for it. If one rejects the idea of there being necessary connections among the property tokens in the world, the consequence is that one has to accept the whole distribution of the fundamental physical properties as primitive. If one shrinks back from that consequence, one has to make a case for there being necessary connections in the world.

## 7.2 Causal properties

One possibility to conceive necessary connections is to admit not only particulars (property tokens), but also universals (property types) and regard the laws of Nature as certain relations among universals (Armstrong, 1983). The laws of Nature are then conceived as governing the development of the distribution of the fundamental physical property tokens in the world. However, on that view, universals are simply added to the Humean account, and the objections against Humeanism can be reiterated on the level of the property types as universals. Furthermore, the ontological relationship between the universals and the particulars remains unclear. I shall therefore not pursue that view in this chapter.

If one remains within the scope of an ontology of property tokens, leaving open whether or not there are universals over and above the property tokens, and nevertheless seeks to recognize metaphysically necessary connections among distinct entities, one has to conceive the particulars that there are in the world as being such that some particulars are the ground of the existence of other particulars – such that, given the former particulars, the latter cannot fail to be there. This is the case if and only if some particulars bring about the existence of other particulars. In other words, they have the power to produce other particulars. Hence, conceiving metaphysically necessary connections among distinct entities within an ontology of particulars commits us to admitting powers. The fundamental physical properties bestow powers on the objects that instantiate these properties or, simply, these properties are powers. I shall use that latter expression for the sake of simplicity (without thereby taking a stance on the issue of the relationship between objects and properties in this chapter). We thus get to a causal view of properties in contrast to the Humean view of properties being categorical. Since the powers in the sense just outlined are fundamental, they are irreducible. They thus are dispositions without a categorical basis – in other words, ungrounded dispositions.

There are two versions of an anti-Humean metaphysics of powers discussed in the current literature.

(1) One version considers each property to be categorical and dispositional in one. More precisely, to the extent that there is a distinction between the categorical and the dispositional, it is a distinction among predicates or descriptions instead of properties (Martin, 1997: in particular sections 3 and 12; Mumford, 1998: chapter 9; Heil, 2003: chapter 11). Each property thus is a power and a quality at the same time.

(2) The other version identifies properties with dispositions in the sense of powers (Shoemaker, 1980; Bird, 2007; as well as Hawthorne, 2001 who calls this view 'causal structuralism' and furthermore Ellis, 2001: in particular chapters 1 and 3; Ellis, however, admits powers as well as categorical properties as two different kinds of properties existing in the world). Properties are powers, consisting in the disposition to produce certain effects.

The difference between these two versions is not as great as it might seem at first glance: the first version does not conceive the distinction between the categorical and the dispositional as an ontological one; it is not even admissible to talk in terms of categorical and dispositional aspects of properties, for such aspects would in turn be properties. Dispositions or powers are not properties of properties, but properties are dispositions or powers. Thus, John Heil describes the last view of Charlie Martin as conceiving properties as 'powerful qualities' (Heil, 2009: 178). The second version does not conceive powers as pure potentialities, but as real, actual properties. They thus are certain qualities that necessarily manifest themselves in the production of certain effects.

The important point is that powers are not additional properties. Qua being a certain way – that is, qua being qualitative –, property tokens are the powers to produce further property tokens. To put the matter in terms of object talk, qua being a certain way, that is, by having certain qualities, objects have the power to produce other objects, or other properties in objects. Take charge for example: insofar as charge is a qualitative property, distinct from e.g. mass, it is the power to build up an electromagnetic field, resulting in the attraction of opposite-charged and the repulsion of like-charged objects. The Humean and the anti-Humean agree about the properties that there are in the world. They disagree about the ontology of properties – whether the property tokens are such that they simply succeed one another or whether they are such that each property token brings about other property tokens.

The metaphysics of powers accepts as primitive that properties are powers – more precisely, accepts as primitive the fact that qua being a certain way, being a certain quality, each property token is the power to produce certain other property tokens. The *laws of Nature* supervene on the properties. If it is a law that all $F$s are followed by $G$s, this is so because the $F$s are the power to produce $G$s. Consequently, instead of the contingent regularity patterns in Humean metaphysics, according to the metaphysics of powers, the laws of Nature are metaphysically necessary: if something is a token of the property $F$, it is the power to produce $G$s. Hence, in any world in which there are $F$s, the law that $F$s produce $G$s holds.

The same goes for *causation*. Causal relations are not a simple pattern of regular contiguous co-instantiations of properties of the same types, but one property token literally produces or brings about other property tokens and thus is the ground of the existence of those other property tokens. However, in recent literature, doubts are expressed as to whether in case one conceives the fundamental properties as dispositions and thus in a causal manner, one inevitably is committed to recognizing necessary connections in the world, thereby contradicting Humean metaphysics (Handfield, 2008; Anjum and Mumford, 2009).

Nonetheless, there is a clear contrast between the Humean metaphysics of categorical properties and a metaphysics that conceives properties in a dispositional and thus a causal manner, namely as powers. According to Humean metaphysics, properties of the same type can stand in very different causal relations in different possible worlds. For instance, Humean metaphysics admits a possible world in which the property type that plays the charge role in the actual world plays the mass role, and *vice versa*. By contrast, according to the metaphysics of powers, the qualitative character of a property type determines the causal role that the property type in question plays and hence determines the causal relations in which the tokens of the property type in question stand. That is why the metaphysics of powers is a causal theory of properties, whereas Humean metaphysics is not a causal theory of properties. Since, following the metaphysics of powers, it is the nature of the properties to produce certain effects, the connection between cause and effect is a necessary one: the property tokens cannot but produce the effects in question.

*Determinism* has a much more substantial meaning in the metaphysics of powers than in Humean metaphysics. The metaphysics of powers suggests the view that if there is an initial state of the world, the big bang, the properties instantiated at the big bang are powers that bring about the subsequent development of the Universe. In other words, the initial state of the world necessitates all there is in the world. Consequently, any possible world that is an exact duplicate of the initial state of the actual world is an exact duplicate of the actual world as a whole. Determinism, thus conceived, clearly raises the problem of whether or not it is compatible with free will. Note that this conception of determinism does not invoke the notion of the laws of Nature. The powers as such account for the world being deterministic. The laws are derived from the powers. Nonetheless, of course, it is true on this account that given the laws and the initial conditions at the origin of the Universe it is in principle possible to predict everything that there is in the Universe.

However, the metaphysics of powers is not committed to determinism. If it were, that would be a weighty objection against it, since there are reasons to believe that the actual world is not deterministic. There are two options within the metaphysics of powers to introduce probabilities: the one possibility is to take the property $F$ to be the power to produce tokens of the property $G$ among others, but to say that it is a matter of pure chance when an $F$ exercises that power. For instance, one may think of two radioactive atoms of the same type in the same environment, the one decays within a given time, the other one does not decay within that time. On this account, however, probabilistic laws are nothing more than Humean regularities: at the end of the world, they simply describe what percentage of the $F$s has produced a $G$.

The other, much more widespread option is to make the notion of a power more precise by employing the concept of propensities (Popper, 1990). On a widespread account of propensities, the power of $F$s to produce $G$s is conceived as a certain tendency intrinsic to each $F$ to bring about a $G$, and that tendency has a certain strength. If, for instance, the strength of the power of $F$s to produce $G$s is 0.6, then that is the reason why 60% of the $F$s in the Universe bring about a $G$. Consequently, probabilistic laws, like deterministic laws, are not Humean regularities. They supervene on the powers, and they are metaphysically necessary. The main advantage of the account in terms of propensities is to provide for objective single case probabilities. Frequencies supervene on the propensities. They are of course the guide to gain cognitive access to the propensities. On that basis, the propensity theory of probabilities can take the Principal Principle into account that adapts our subjective degrees of beliefs to objective chance: in finding out what the correct probability distribution is in a given case, we learn which subjective degrees of belief are rational to adopt.

The metaphysics of causation in terms of powers is linked with the view of an intrinsic direction of time. The connection between a token of $F$ and a token of $G$ is a causal one if and only if the $F$ produces the $G$. One can argue that causation thus is the basis for the direction of time: the $G$, being produced by the $F$, lies in the future of the $F$. Causal processes are the first and foremost example of irreversible processes: the relationship between a cause and its effect cannot be reversed. There hence is a direction of time because there are causal processes.

As regards the initial state of the Universe, it is simply a contingent matter of fact that, according to the past hypothesis, this is a state of extremely low entropy. Whereas the Ramsey–Lewis view of laws can promote the description of that state to a theorem of the best system, the metaphysics of powers simply has to acknowledge the low entropy of the initial state of the Universe to be a contingent fact. Nonetheless, any metaphysical position has to accept something as primitive. In contrast to Humean metaphysics, the metaphysics of powers does not have to regard the whole distribution of fundamental physical properties in space-time as primitive. Powers establish a necessary connection between at least some of these properties. In the ideal case of a completely deterministic world, one only has to accept the initial state of the Universe as primitive, and the powers instantiated at the initial state of the Universe are the ground of the existence of the development of the distribution of the fundamental physical properties throughout the whole of space-time.

To sum up, the issue of Humean metaphysics vs. a metaphysics of powers concerns the categorical vs. the causal view of properties. If the fundamental physical properties are categorical, then the whole distribution of them has to be accepted as primitive, because there are no necessary connections among them. If, on the contrary, the fundamental physical properties are powers, then a proper part of the property tokens instantiated in the world (in the ideal case, those ones at the big bang) is the ground of the existence of the rest of the fundamental property tokens in the world, there thus being necessary connections between property tokens in the world.

## 7.3 The master argument against Humeanism and the ontology of science

There is a central objection that can be seen as the master argument against Humeanism. According to Humean metaphysics, the causal relations and the laws of Nature vary from possible world to possible world, depending on what else there is in a world in which properties of the type $F$ are instantiated. Thus, what $F$ is neither depends on the effects that the tokens of $F$ have nor on the laws in which $F$ figures in different possible worlds. The objection now is this one: we gain knowledge about the world via the causal relations in which what there is in the world stands. A difference that does not give rise to a causal difference is a difference of which we cannot gain knowledge. Hence, we can specify what there is in the world only down to causal equivalence. If the properties themselves are not causal ones, but categorical, intrinsic ones, possessing an essence beyond that specification, it follows that their essence is a primitive suchness (quiddity) that is beyond the grasp of our knowledge. The Humean thus is committed to the positions known as quidditism and as humility (Lewis, 2009 endorses both these commitments).

The commitment to quidditism and humility as its consequence are rather uncomfortable: *there are worlds that are different because they differ in the distribution of the intrinsic properties that are instantiated in them, although there is no difference in causal and nomological relations and thus no discernible difference between them*. A quidditistic difference between worlds is a qualitative difference (by contrast to a haecceistic difference, concerning the question which individuals there are in a given world) that implies that worlds have to be counted as different although they are indiscernible (Black, 2000). A gap between metaphysics – postulating primitive qualities (a primitive suchness) – and epistemology thus arises. These commitments are at odds with a metaphysics that seeks to be close to empirical science – postulating that there are primitive qualities in the world that lead to having to recognize possible worlds as qualitatively different although they are indiscernible. These consequences therefore constitute a strong argument in favour of the causal theory of properties in general (Shoemaker, 1980 and 2007: appendix).

Hence, although Humean metaphysics seems to be parsimonious at first glance, it is not that parsimonious after all, postulating that properties have an essence beyond the causal relations into which they enter, that essence consisting in a primitive suchness which is beyond the grasp of our knowledge. Deleting these commitments leads to the causal theory of properties, namely to the view that there is nothing more to the properties than the power to bring about certain effects, and that view then commits us to countenance necessary connections, as explained above. However, seen from the perspective of avoiding the commitments to quidditism and humility, the commitment to necessary connections is not at all mysterious or ontologically inflationary.

If we take properties to be causal powers and if the main argument for adopting that view is to abandon the commitments to quidditism and humility, then the fundamental properties cannot be conceived as needing outside manifestation conditions for exercising the powers that they are. Otherwise, there could be fundamental properties of two different types $F$ and $F^*$ present in the world without that difference showing up anywhere because the appropriate manifestation conditions always lack. It is unproblematic to conceive the fundamental properties as not needing outside manifestation conditions in order to exercise the powers that they are, since these powers are not pure potencies, but real, actual properties. The appropriate model are not macroscopic dispositions such as the disposition of water to dissolve sugar, but rather the dispositions of radioactive atoms for spontaneous decay or of particles with charge to build up electromagnetic fields.

However, it seems questionable whether the causal theory of properties is compatible with physics. Since Russell's 1912 famous paper denouncing the notion of causation as production, a lot of Humeans claim that the notion of powers does not fit into our fundamental physical theories. All that these theories state are certain regularities in the distribution of physical properties. There is thus not more to causation than Humean regularities (cf. notably the papers in Price and Corry, 2007).

Nonetheless, the situation is not as clear as Russell and his contemporary followers represent it. Characterizations of the fundamental physical properties in dispositional terms are widespread (Mumford, 2006: 475–477). As regards the fundamental physical, intrinsic properties, one can conceive charge, for instance, as the power to build up an electromagnetic field, mass as the power to resist acceleration, etc. It is true that the causal conception of these properties is not mandatory. One can also characterize these properties in terms of certain mathematical structures, namely as being invariant under certain symmetry relations that constitute a mathematical group (Psillos, 2006: 151–154). However, it is trivial that any physical property can be represented by a mathematical structure. That representation does not imply that the property itself is a mathematical structure, on pain of blurring the distinction between the mathematical and the physical. In short, that manner of representation leaves open what the nature of the properties thus represented is. As far as fundamental physical, intrinsic properties such as charge or mass are concerned, instead of taking their essence to be a primitive suchness (quiddity), it is an attractive and viable option in the philosophy of physics to consider them as causal powers.

In the recent literature in the philosophy of physics, stress is laid not only on mathematical structures representing physical properties, but it is claimed with sound arguments from the current fundamental theories (quantum theory and general relativity theory) that some central fundamental physical properties are themselves physical structures rather than being intrinsic properties. A physical structure can be conceived as a network of concrete qualitative, physical relations among objects that do not possess any intrinsic identity and thus no intrinsic properties on which these relations could supervene. The resulting position is known as ontic structural realism (Ladyman, 1998; French and Ladyman, 2003; Esfeld, 2004). However, the founders of ontic structural realism tend to conceive these physical structures in an anti-Humean way, namely as including a primitive modality (French, 2006; Ladyman and Ross, 2007: chapters 2–5). One way of elaborating on that view of a primitive modality is to apply the causal theory of properties to the fundamental physical structures as conceived by ontic structural realism.

As far as quantum theory is concerned, the argument in support of ontic structural realism is that the state-dependent properties of quantum objects are subject to entanglement such that, in brief, there are only relations of entanglement among these objects and no state-dependent properties that each of these objects possesses on its own. State-dependent properties are position, momentum, spin angular momentum in each of the three spatial directions, etc. If one recognizes entangled states and if one maintains that there also are classical physical properties, at least when it comes to macroscopic objects, one has to conceive a transition from entangled states to so-called product states, which amount to state-dependent properties of quantum objects with definite numerical values. The most elaborate physical proposal for such a transition is the one going back to Ghirardi *et al.* (1986) (GRW). The interpretation of quantum theory by GRW lends itself to an account in terms of dispositions: the entangled states are the power – more precisely, the propensity – to produce product states, that is, classical physical properties with definite numerical values localized in classical space-time (Dorato, 2006; Suárez, 2007). Conceiving the entangled states in that manner provides for a clear answer to the question what the properties of quantum objects are if there are no properties with definite numerical values, namely propensities for spontaneous localizations. Furthermore, it yields objective single case probabilities and it makes clear how a structure can be a power and how it can exercise the power that it is without needing external manifestation or triggering conditions: the disposition (propensity) in question is one for *spontaneous* localization. Since entangled states are nonseparable, it is evident that this disposition (propensity) has to be inherent in the entangled state as a whole, viz. that the entangled state as a whole *is* the disposition or power for spontaneous localization.

As far as general relativity theory is concerned, there are sound arguments, notably the hole argument, for conceiving the metrical field as being part and parcel of space-time and thus as regarding the properties of space-time points as consisting in the metrical relations among them (Esfeld and Lam, 2008). If the metrical field belongs to space-time, space-time is no passive background arena, but itself a dynamical entity, containing energy, namely the gravitational energy, and gravitation is a physical interaction on a par with the other physical interactions, such as electromagnetism. This situation makes it possible not only to apply ontic structural realism to space-time as well, but also to conceive the metrical relations themselves in a causal manner, namely as the power to produce the observed gravitational phenomena among others (Bartels, 1996: 37–38; Bartels 2009; Bird 2009: section 2.3). One can thus avoid an unsatisfactory dualism of on the one hand causal properties or structures and on the other hand purely categorical, structural properties.

Again, it is not mandatory to conceive the fundamental physical structures in a causal manner as powers. But there are mainly three arguments for doing so: (1) Applying the causal theory of properties to the fundamental physical structures provides for a clear answer to the question what distinguishes physical from mathematical structures: physical structures are causal, whereas mathematical structures are not. (2) This conception yields a unified ontology that includes properties such as charge and mass, that is, physical properties for which there is no argument available that is comparable to the arguments from state-dependent properties in entangled states and metrical properties and that shows that these properties themselves are – physical – structures. (3) This conception establishes a firm link between the fundamental physical structures, which are theoretical entities, and the observable phenomena, making clear how we can have cognitive access to these structures and thus opening up the way for vindicating a realist attitude towards them (Esfeld, 2009).

The last argument calls into question whether Humean metaphysics really is entitled to realism with respect to the fundamental physical properties. Whereas Lewis himself considers these properties to be intrinsic, Humean metaphysics can be adapted to current physics by shifting the focus from intrinsic properties to relations or structures (such as the structures of quantum entanglement) (Sparber, 2009). The point at issue, however, is whether Humean metaphysics can justify a realist stance with respect to the theoretical entities in which current physics trades: since, according to the Humean, the nature of these entities does not consist in bringing about effects that are observable, what the underlying structures are seems to be underdetermined by the observable effects. Consequently, the result risks to be an agnosticism rather than a scientific realism with respect to the fundamental physical structures (cf. the empiricist structural realism of van Fraassen, 2006). Such a realism, however, is crucial for Humean metaphysics: Lewis maintains in his thesis of Humean supervenience quoted above that all there is to the world is the distribution of the fundamental physical, categorical properties in space-time. Everything that there is in the world is a feature of that distribution. Thus, everything is traced back to the theoretical entities of fundamental physics.

Nonetheless, there is more to science than physics. There also are the non-physical special sciences, such as notably biology and psychology. These sciences trade in functional properties. Those properties consist in producing certain effects given standard conditions. Thus, for instance, a gene consists in producing certain specific phenotypical effects given standard conditions; a functional, mental property consists in producing certain further mental properties as well as a certain behaviour given standard conditions, etc. Functional properties, however, cannot be conceived as categorical properties, for what they are consists in the effects that they cause. What these properties are can thus not be separated from the causal and the nomological relations in which they stand. Humean metaphysics can acknowledge functional descriptions: the distribution of the categorical fundamental physical properties (or structures) over the whole Universe can be such that it makes true certain functional descriptions. But Humean metaphysics cannot admit functional properties: in a Hume world, there cannot be properties whose identity consists in causing certain effects, for all there is in a Hume world is the distribution of categorical properties. Hence, Humean metaphysics may be in the position to provide truthmakers for the functional descriptions and theories of the special sciences, but it is committed to an eliminativist attitude with respect to the functional properties in which these sciences trade (Esfeld, 2007).

In contrast, the metaphysics of powers has no problem in accommodating the functional properties in which the special sciences trade. One may go as far as maintaining that the causal theory of properties (Shoemaker, 1980) is tailor-made for functional properties, since what a property is consists in the effects that the property tokens in question produce. However, as mentioned above, there are also good arguments for extending the causal theory of properties to the physical properties, including the fundamental physical properties and structures. The metaphysics of powers thus is an ontology that pays tribute to the commitments of both physics and the special sciences. As far as that ontology is concerned, nothing hinders that the properties in which the special sciences trade are at least token-identical with certain physical properties. But that identity is in any case conservative instead of an eliminativism with respect to the special sciences' properties, since what the properties are consists in the production of certain effects, and some of these effects are those ones on which the special sciences focus.

In conclusion, summing up this section, there are two main arguments for preferring the metaphysics of powers to Humean metaphysics. (1) The metaphysics of powers avoids the commitments to quidditism and humility. (2) It provides for a complete and coherent ontology that includes the ontological commitments of physics as well as of the special sciences.

## References

Anjum, R. L. and Mumford, S. (2009). Dispositional modality. In *Lebenswelt und Wissenschaft. XXI. Deutscher Kongress für Philosophie. Kolloquien*, ed. C. F. Gethmann, Hamburg: Meiner, in press.

Armstrong, D. M. (1983). *What is a Law of Nature?* Cambridge: Cambridge University Press.

Bartels, A. (1996). Modern essentialism and the problem of individuation of spacetime points. *Erkenntnis*, **45**, 25–43.

Bartels, A. (2009). Dispositionen in Raumzeit-Theorien. In *Lebenswelt und Wissenschaft. XXI. Deutscher Kongress für Philosophie. Kolloquien*, ed. C. F. Gethmann, Hamburg: Meiner, in press.

Beebee, H. (2006). Does anything hold the world together? *Synthese*, **149**, 509–533.
Beebee, H. and Mele, A. (2002). Humean compatibilism. *Mind*, **111**, 201–223.
Bird, A. (1998). Dispositions and antidotes. *Philosophical Quarterly*, **48**, 227–234.
Bird, A. (2007). *Nature's Metaphysics. Laws and Properties*. Oxford: Oxford University Press.
Bird, A. (2009). Structural properties revisited. In *Dispositions and Causes*, ed. T. Handfield. Oxford: Oxford University Press, pp. 215–241.
Black, R. (2000). Against quidditism. *Australasian Journal of Philosophy*, **78**, 87–104.
Callender, C. (2001). Thermodynamic asymmetry in time. In *Stanford Encyclopedia of Philosophy*, ed. E. N. Zalta. http://plato.stanford.edu/archives/win2001/entries/time-thermo.
Choi, S. (2008). Dispositional properties and counterfactual conditionals. *Mind*, **117**, 795–841.
Cross, T. (2005). What is a disposition? *Synthese*, **144**, 321–341.
Dorato, M. (2006). Properties and dispositions: some metaphysical remarks on quantum ontology. In *Quantum Mechanics: Are There Quantum Jumps? On the Present State of Quantum Mechanics, American Institute of Physics Conference Proceedings*, vol. ed. A. Bassi, D. Dürr, T. Weber and N. Zanghi. New York: American Institute of Physics, pp. 139–157.
Ellis, B. (2001). *Scientific Essentialism*. Cambridge: Cambridge University Press.
Esfeld, M. (2004). Quantum entanglement and a metaphysics of relations. *Studies in History and Philosophy of Modern Physics*, **35B**, 601–617.
Esfeld, M. (2007). Mental causation and the metaphysics of causation. *Erkenntnis*, **67**, 207–220.
Esfeld, M. (2009). The modal nature of structures in moderate structural realism. *International Studies in the Philosophy of Science*, **23**, 179–194.
Esfeld, M. and Lam, V. (2008). Moderate structural realism about space-time. *Synthese*, **160**, 27–46.
French, S. (2006). Structure as a weapon of the realist. *Proceedings of the Aristotelian Society*, **106**, 167–185.
French, S. and Ladyman, J. (2003). Remodelling structural realism: quantum physics and the metaphysics of structure. *Synthese*, **136**, 31–56.
Frigg, R. and Hoefer, C. (2007). Probability in GRW theory. *Studies in History and Philosophy of Modern Physics*, **38B**, 371–389.
Ghirardi, G., Rimini, A. and Weber, T. (1986). Unified dynamics for microscopic and macroscopic systems. *Physical Review D*, **34**, 470–491.
Gundersen, L. (2002). In defense of the conditional-account of dispositions. *Synthese*, **130**, 389–411.
Handfield, T. (2008). Humean dispositionalism. *Australasian Journal of Philosophy*, **86**, 113–126.
Hawthorne, J. (2001). Causal structuralism. *Philosophical Perspectives*, **15**, 361–378.
Heil, J. (2003). *From an Ontological Point of View*. Oxford: Oxford University Press.
Heil, J. (2009). Obituary. C. B. Martin. *Australasian Journal of Philosophy*, **87**, 177–179.
Ladyman, J. (1998). What is structural realism? *Studies in History and Philosophy of Modern Science*, **29**, 409–424.
Ladyman, J. and Ross, D. (2007). *Every Thing Must Go: Metaphysics Naturalised*. Oxford: Oxford University Press.
Lewis, D. (1973a). Causation. *Journal of Philosophy* **70**, 556–567. Reprinted in D. Lewis (1986). *Philosophical Papers*. vol. 2. Oxford: Oxford University Press, pp. 159–172.
Lewis, D. (1973b). *Counterfactuals*. Oxford: Blackwell.

Lewis, D. (1986). *Philosophical Papers*, vol. 2. Oxford: Oxford University Press.
Lewis, D. (1994). Humean supervenience debugged. *Mind*, **103**, 473–490. Reprinted in D. Lewis (1999). *Papers in Metaphysics and Epistemology*. Cambridge: Cambridge University Press, pp. 224–247.
Lewis, D. (1997). Finkish dispositions. *Philosophical Quarterly*, **47**, 145–158. Reprinted in D. Lewis (1999). *Papers in Metaphysics and Epistemology*. Cambridge: Cambridge University Press, pp. 133–151.
Lewis, D. (2004). Causation as influence. In *Causation and Counterfactuals*, ed. J. Collins, N. Hall and L. A. Paul. Cambridge, MA: MIT Press, pp. 75–106.
Lewis, D. (2009). Ramseyan humility. In *Conceptual Analysis and Philosophical Naturalism*, ed. D. Braddon-Mitchell and R. Nola. Cambridge, MA: MIT Press, pp. 203–222.
Loewer, B. (2007). Counterfactuals and the second law. In *Causation, Physics, and the Constitution of Reality. Russell's Republic Revisited*, ed. H. Price and R. Corry. Oxford: Oxford University Press, pp. 293–326.
Malzkorn, W. (2000). Realism, functionalism, and the conditional analysis of dispositions. *Philosophical Quarterly*, **50**, 452–469.
Martin, C. B. (1994). Dispositions and conditionals. *Philosophical Quarterly*, **44**, 1–8.
Martin, C. B. (1997). On the need for properties: the road to Pythagoreanism and back. *Synthese*, **112**, 193–231.
Mumford, S. (1998). *Dispositions*. Oxford: Oxford University Press.
Mumford, S. (2006). The ungrounded argument. *Synthese*, **149**, 471–489.
Popper, K. R. (1990). *A World of Propensities*. Bristol: Thoemmes.
Price, H. and Corry, R. (2007). *Causation, Physics, and the Constitution of Reality. Russell's Republic Revisited*. Oxford: Oxford University Press.
Psillos, S. (2006). What do powers do when they are not manifested? *Philosophy and Phenomenological Research*, **72**, 137–156.
Russell, B. (1912). On the notion of cause. *Proceedings of the Aristotelian Society*, **13**, 1–26.
Shoemaker, S. (1980). Causality and properties. In *Time and Cause*, ed. P. van Inwagen. Dordrecht: Reidel, pp. 109–135. Reprinted in S. Shoemaker (1984). *Identity, Cause, and Mind. Philosophical Essays*. Cambridge: Cambridge University Press, pp. 206–233.
Shoemaker, S. (2007). *Physical realization*. Oxford: Oxford University Press.
Sparber, G. (2006). Powerful causation. In *John Heil. Symposium on his Ontological Point of View*, ed. M. Esfeld. Frankfurt: Ontos, pp. 123–137.
Sparber, G. (2009). *Unorthodox Humeanism*. Frankfurt: Ontos-Verlag.
Suárez, M. (2007). Quantum propensities. *Studies in History and Philosophy of Modern Physics*, **38B**, 418–438.
van Fraassen, B. C. (2006). Structure: its shadow and substance. *British Journal for the Philosophy of Science*, **57**, 275–307.

# Part III

Reduction

# 8

# The crystallization of Clausius's phenomenological thermodynamics

C. ULISES MOULINES

## 8.1 Introduction

The (metatheoretical) structuralist program for the reconstruction of scientific theories (in the following referred to as 'structuralism') allows for the performance of several metatheoretical tasks, among them: (a) precisely explicating the inner structure of any given 'textbook' theory in all its complexity; (b) making a clear distinction between different theories that in standard expositions are usually not clearly distinguished from each other, and identifying their intertheoretical relationships; (c) reconstructing their evolution in historical time. The present essay applies the structuralist methodology to the analysis of two writings of the German physicist Rudolf J. Clausius on thermodynamic phenomena published in the middle of the nineteenth century. Conventional historiography interprets these writings as the presentation of a first well-founded theory of phenomenological thermodynamics. However, as our analysis will show, Clausius actually constructed in his papers *three* different theories – each of them with its own identity, though of course related to each other.

## 8.2 The essentials of the structuralist representation of theories

Structuralism (without this name) was initiated by the pioneering works of Joseph D. Sneed, *The Logical Structure of Mathematical Physics* (1971) and Wolfgang Stegmüller, *Theorienstrukturen und Theoriendynamik* (1973). However, the present chapter employs the more evolved conceptual tools and representation methods to be found in the joint work, *An Architectonic for Science*, by Wolfgang Balzer, C. Ulises Moulines, and Joseph D. Sneed (1987). For those readers who are not well-acquainted with structuralist ideas, a brief summary of them is laid out in what follows.

According to structuralism, the term 'theory' is polisemic: Its ordinary use denotes at least three different kinds of entities, which differ according to three different levels of aggregation: *theory-elements*, *theory-nets*, and *theory-holons*. A theory-net is a hierarchically organized array of theory-elements with a 'common basis' and a common conceptual framework; and a theory-holon is a complex of theory-nets with different conceptual structures related to each other by certain intertheoretical links. Since the notion of a theory-holon will play no role in the present analysis, we can skip it out and concentrate on theory-elements and theory-nets. The 'essential identity' of a theory-element is given by two constituents: a homogeneous class of models and a domain of intended applications described by partial structures of the models (their relative non-theoretical parts). (Besides these two constituents, according to structuralism, a 'normal' theory-element is also constituted by other, more complex model-theoretic structures; nevertheless, for the purposes of the present chapter, we may forget about them.)

A model is a structure of the form

$$\langle D_1, \ldots, D_m; A_1, \ldots, A_n; R_1, \ldots, R_p; f_1, \ldots, f_q \rangle,$$

where the $D_i$ are the theory's basic domains (its 'ontology'), the $A_i$ are auxiliary sets of mathematical entities (usually numbers), the $R_i$ are relations on (some of) the basic sets, and the $f_i$ are (metrical) functions on (some of) the basic and auxiliary sets.

It is not very important how we determine these models – whether by thoroughly formal means, or semi-formally, or in ordinary language. But it seems that the most practical method is to define the class of a theory's models by means of the Suppesian method of defining a so-called *set-theoretical predicate*, i.e. a predicate defined solely by means of the tools of naive or semi-formal set theory. (We will see numerous examples of such predicates when we come to the reconstruction of Clausius's theories.)

A homogeneous class of models, **M**, is a class of structures satisfying all the same set-theoretical predicate. In empirical theories, to this class of models a domain of intended interpretations, **I**, is associated: This is a class of substructures of *some* of the elements of **M**, containing (almost) all the $D_i$ and $A_i$, but only some of the $R_i$ and $f_i$ – those that are non-theoretical with respect to the theory in question, i.e. those whose values may be determined without presupposing the substantial laws used to define the set-theoretical predicate which, in turn, determines **M**. The other relations and functions, if any, are called 'relatively theoretical' or '*T*-theoretical'. The domain of intended applications of a given theory represents the physical systems to which the theory is supposed to apply, where it has to be tested, or in other words, its empirical basis.

Though the distinction between *T*-theoretical and *T*-non-theoretical notions is an important constituent of the structuralist metatheory, its application to concrete case-studies is usually a quite intricate issue, which involves a lengthy and careful discussion. For lack of space, I will not go into this issue when analysing Clausius's theories here. I will only make some brief remarks about those of his notions that clearly seem to be 'Clausius-theoretical' according to the reconstruction proposed.

Thus, a theory-element is (in this highly simplified version of structuralist methodology) an ordered pair $\langle \mathbf{M}, \mathbf{I} \rangle$.

A theory-net is a finite class of theory-elements ordered by a relation of *specialization* and having a first element. This first element is what may be called 'the *basic theory-element*', $\mathbf{T}_0$ ($= \langle \mathbf{M}_0, \mathbf{I}_0 \rangle$). It is the one where the models are determined only by the fundamental law(s) of the theory in question. Taken in isolation, the fundamental law(s) in $\mathbf{T}_0$ usually have an empirically almost vacuous nature – they are 'guiding principles' rather than statements of fact. But for this very reason, they constitute the theory's core. All other theory-elements $\mathbf{T}_i$ of the net come out of $\mathbf{T}_0$ by a process of specialization, i.e. by adding more special (and more empirical) laws and conditions to the fundamental law(s), and by restricting the range of intended applications. Clearly, for any $i$, $\mathbf{M}_i \subseteq \mathbf{M}_0$. Typically, a theory-net has the form of an inverted tree:

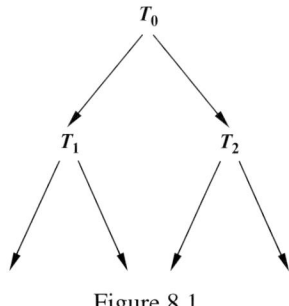

Figure 8.1

$\mathbf{T}_1$, $\mathbf{T}_2$, and so on are specializations of $\mathbf{T}_0$.

Up to this point, a purely *synchronic* point of view with respect to the structure of theories has been taken. But the structuralist methodology may be used to represent also the *diachronic* structure of science. As a matter of empirical (historical) fact, it is assumed that at least *four main types* of diachronic processes may be distinguished in the history of science (see Moulines, 1996). All of them can be, and have been, formally represented within the structuralist framework. These are:

- *Evolution*: A theory-net evolves in time, adding or suppressing some specialized theory-elements but without losing its 'essential identity', which is given by $\mathbf{T}_0$. This roughly corresponds to what Thomas S. Kuhn called 'normal science' and Imre Lakatos 'a research program'. Example: the evolution of Newtonian mechanics from the end of the seventeenth century to the beginning of the twentieth century (see Balzer *et al.* 1987: Ch. V).

- *Replacement*: A theory aimed at systematizing a given domain of intended applications is completely replaced by another theory (with a very different conceptual framework) aimed at more or less the same domain of intended applications. This corresponds to one sense of what Kuhn has called a 'scientific revolution'. Example: the replacement of phlogiston theory by the oxidation theory (see Caamaño, 2004).
- *Embedding*: The models of a previous theory are (approximatively, and perhaps not completely) embedded into the models of a more complex theory. This roughly corresponds to another sense of Kuhnian revolutions. Example: the embedding of Keplerian planetary theory into Newtonian mechanics (see Balzer *et al.* 1987: Ch. VII).
- *Crystallization*: After the breakdown of a previous theory, and through a long and gradual process, the models of a new theory are constructed in a piecemeal fashion, with many intermediate, fragmentary states before a fully developed new theory-net appears. Such a process is neither 'normal' nor 'revolutionary' in anything like Kuhn's sense, and I know of no other philosophers of science that have dealt with this kind of process (with the possible exception of Elkana, 1974). But I contend that it is quite frequent in the history of science. Crystallization is much more complex than the previously mentioned types of diachronic structures; it contains some elements of the other types but it also has its own features. The slow emergence of phenomenological thermodynamics after the breakdown of the caloric theory seems to be a clear case in point. In this process, the work of Clausius analysed in this paper played a crucial role (though it was only a part of the whole process).

## 8.3 The historical data

The 'paradigm' for the theory of heat from the end of the eighteenth century until the mid-1820s was the caloric theory developed by Lavoisier, Laplace, Biot, and others. This was a quite successful theory until it began to agonize in the late 1820s due to 'internal' (conceptual) as well as 'external' (application-related) 'anomalies'. A new 'paradigm' for the theory of heat, i.e. a full-fledged theory which was fully satisfactory both from the conceptual-mathematical and from the empirical point of view was to appear only much later, in 1876, with J. Willard Gibbs's *On the Equilibrium of Heterogeneous Substances* (1st part). There we have for the first time what we now call 'phenomenological thermodynamics' in its presently acknowledged form, almost indistinguishable from what contemporary textbooks on thermodynamics lay out (see, for example, Callen, 1960). In the meantime, that is, between the 1820s and the 1870s, we perceive a long and chaotic period of partial attempts at the construction of a fully satisfactory theory of heat, which were partially successful but also partially failed, with very different theoretical concepts and principles. In this long process, the names of Carnot, Clapeyron,

Regnault, Mayer, Helmholtz, Joule, Kelvin, Rankine and of course Clausius, among others, play an important role, all of them working from different standpoints and obtaining different, sometimes incomparable results. The culmination of this process was Gibbsian thermodynamics (see Moulines, 1991). Clausius's work was an essential part of this process, itself a fragmentary crystallization within the general crystallization process. This is what is to be reconstructed in the next pages.

In the 1850s, Clausius published three papers on the theory of heat: *Über die bewegende Kraft der Wärme* (1850), *Über eine veränderte Form des zweiten Hauptsatzes der mechanischen Wärmetheorie* (1854) and *Über die Art der Bewegung, die wir Wärme nennen* (1857). The last paper amounts to the first consistent exposition of the kinetic theory of gases, and in this sense, it is the forerunner of statistical mechanics; as such, it is of no concern to us here, since we are dealing with the development of (phenomenological) thermodynamics – which is what Clausius, and others, called then the *Mechanische Wärmetheorie* – the 'mechanical theory of heat'. The three papers taken together might be said to constitute a part of a theory-holon. But since we are not going to deal with such complex structures here, let us restrict our attention to the papers of 1850 and 1854. The 1850 essay contains one theory, while the 1854 paper contains *two* theories. Each one of these theories may consistently be represented as a theory-net in our sense. To abbreviate, we will denote the 1850-theory(-net) by '**Cl-1**', while the two theories (i.e. theory-nets) laid out in the 1854 paper will be named '**Cl-2.1**' and '**Cl-2.2**', respectively.

## 8.4 The basic theory-element of Cl-1

The 1850 paper is divided in two parts, respectively entitled: 'I. Consequences of the principle of the equivalence of heat and work', and 'II. Consequences of Carnot's principle in connection with the former'.[1] Given the form of their presentation, one could be misled in thinking that Clausius lays out *two* different fundamental principles: the principle of the heat/work equivalence, and the hypothesis that Clausius considers to be the core of the so-called 'Carnot principle', namely that 'whenever work produces heat while not producing a permanent transformation in the state of the working body, a certain amount of heat goes from the hot body to a cold one' (Clausius, 1850: 4). Now, this interpretation, which is still to be found in contemporary expositions, seems to be misleading, both from the point of view of Clausius's original text and of a coherent formal reconstruction of them. In fact, Clausius himself stresses several times that both principles are 'intimately

---

[1] Here, as in the rest of the chapter, I translate as truthfully as I can from the original German text. The page numbers indicated in the quotations are those of the original essay.

linked'. It is more adequate, as we shall see, to interpret both 'principles' as two specifications of more general and fundamental principle that Clausius takes implicitly as his starting point.

For the structuralist reconstruction of **Cl-1**, the first question to be asked is what its 'basic ontology' is, that is what kinds of objects, and what magnitudes defined over those objects we need to formulate the theory's laws.

(1) Clausius speaks once and again of *bodies*, that is, of concrete, spatio-temporally located entities that may be in physical contact and/or have component parts. In some passages of his paper, he also has to speak about a more abstract kind of entities: *chemical substances* to which the concrete bodies may belong; but this he does only for some specializations of his theory. This means that his basic 'ontological commitment' is only with bodies, whereas the 'commitment' with *kinds* of bodies only occurs in some of the theory's specializations (in particular, for the treatment of ideal gases). Also, Clausius speaks of bodies as being in a certain *state*, or as changing their state, though he does not specify what these 'states' might be. This is all that is to be said about the ontology of **Cl-1**. Consequently, the basic sets of each model of **Cl-1** will be:
- A finite set of bodies: $K = \{k_1, k_2, \ldots\}$;
- A continuous sequence of states: $Z \cong \Re+$.[2]

(2) Clausius needs a certain number of primitive magnitudes to formulate his laws. From a formal point of view, the problem with **Cl-1** is that this theory deals with *two* quite different kinds of intended applications: ideal gases and steam at a maximum density. The special laws relevant for one and the other type of application contain indeed some common magnitudes, but other magnitudes are specific either of ideal gases or of steam. Formally, this means that in the theory-net representing **Cl-1**, the models of the basic element will contain some magnitudes common to all theory-elements of the net, while in the specializations some further magnitudes have to be added as primitive concepts – the models becoming thereby more complex structures.

The magnitudes common to all elements of the theory-net are:

$A$ (a universal constant settling the heat/work equivalence);[3]
$P$ (pressure);
$V$ (volume);
$t$ ('empirical' temperature);
$Q$ ('heat');
$W$ ('work').

There is no doubt that $Q$ and $W$ are basic concepts in **Cl-1** – whatever usual textbooks and historical expositions, according to which 'heat' was not a fundamental

---

[2] One needs a continuous sequence of states as arguments of differentiable magnitudes. Clausius assumes throughout his text without further ado that most thermodynamic magnitudes are differentiable functions.
[3] Clausius assumes that, according to Joule's experiments, the value of $A$ should be approximately 1/421. But $A$'s concrete value is irrelevant for the formal reconstruction of the theory.

notion for Clausius anymore, might claim. He speaks all the time about these magnitudes, they explicitly appear in his 'fundamental principles' as well as in the more special laws, and they are supposed to be differentiable functions. (Though Clausius does not specify with respect to what kind of parameters they are differentiable, it is clear from the context that these are the states a body goes through.) On the other hand, it is also true that Clausius substantially departs from previous approaches to heat and work (essentially those of the caloric theory), especially by his explicit denial that $Q$ is a conservative magnitude and by 'identifying' it with $W$. In fact, the discussion at the beginning of his essay seems to make clear that a new interpretation of $Q$ and $W$ is needed, whereby these notions only make sense if one accepts the 'principle of equivalence' between $Q$ and $W$. In structuralist terms, this means that, *in this theory*, $Q$ and $W$ are to be regarded as **Cl-1**-theoretical.

The specific concepts to be *added* in order to formulate the specializations are, respectively,

- for the ideal gases:

    $\Gamma$ (a finite set of gaseous substances – this notion is implicit in the original text);

    $R$ (a parameter specific of each gaseous substance which depends on the substance's specific weight; therefore, it is not identical with *our* universal constant '$R$' but it is related to it);

    $c_V$ (specific heat at constant volume);

    $c_P$ (specific heat at constant pressure);

    $U$ (an 'anonymous' or 'arbitrary' function – '*willkürlich*' is the German word Clausius employs here; it is undoubtedly what later will be called 'internal energy', appearing in this text for the first time in the history of thermodynamics);[4]

    $a$ (a constant specific of ideal gases);[5]

- for steam at a density maximum:

    $\Sigma$ (a finite set of substances with a liquid and a steam phase – implicit in the original text);

    $s$ (steam volume at a maximum density);

    $\varsigma$ (liquid volume at a maximum density);

    $m$ (mass of liquid transformed into steam);

    $c$ (the liquid's specific heat);

    $h$ (another 'anonymous' function required to formulate certain differential equations, of which Clausius only says that it should depend on the steam's temperature).

---

[4] A further interesting aspect of $U$ should be noted: Clausius explicitly avows that he introduces this magnitude in order to formulate a differential equation that will allow for a derivation of the ideal gas law, $U$ making sense only in the context of the equation at stake. In other words, $U$ appears to be **Cl-1**-theoretical. The same goes for the function $h$ introduced later on – see below.

[5] Clausius writes at the end of his paper that experimental results suggest that the actual value of $a$ should lie 'around 273'; but, again, the concrete value is unimportant for the formal reconstruction.

Having settled the 'ontology' and the conceptual framework of **Cl-1**, the next step is to provide an adequate reconstruction of the fundamental law(s) of the basic theory-element of **Cl-1**. A general metatheoretical hypothesis of structuralism is that, in theories having attained a certain degree of unity and structural complexity, there will be *just one* fundamental law, playing the role of a guiding principle and having (almost) no empirical content when taken in isolation. At first sight, **Cl-1** seems to contravene this hypothesis, since Clausius explicitly states two axioms: the Q/W-equivalence and 'Carnot's Principle'. However, I will argue that appearances mislead us here, and that the metatheoretical just-one-fundamental-law hypothesis continues to be valid in the present case.

Clausius formulates the Q/W-equivalence principle in the first part of his article:

the heat consumed divided by the work produced = A.

*(Clausius, 1850: 17)*

As for the Carnot Principle, which is relevant for the second part of the essay, Clausius states it in a purely informal way:

to the production of work as equivalent there corresponds a mere transfer of heat from a hot body to a cold one.

*(Clausius, 1850: 30)*

It is striking that, when Clausius goes on to formulate the Carnot Principle in the second part of his article, he *presupposes* that there is an equivalence of work and heat, that is, he presupposes the first principle. And, as already remarked, there are other passages in the article where he emphasizes that both principles are 'intimately connected'. Therefore, we are confronted with two possible interpretations: (a) the equivalence principle is the theory's only fundamental law and the Carnot Principle is a *specialization* thereof in the structuralist sense; or (b) both principles constitute in fact *two aspects* of one and the same (implicit) guiding principle that is to be formulated synthetically. Both interpretations are coherent with the rest of the original text. Nevertheless, I favour the second one since it is easy to make explicit and, moreover, it seems to me to be more in accordance with Clausius's original intentions. Indeed, if we admit that $Q$ and $W$ are differentiable functions (as Clausius himself presupposes all the time), then it is immediate to state both principles formally as:

$$\text{Equivalence principle} : \frac{dQ(k,z)}{dW(k,z)} = A \in \Re$$
$$\text{Carnot's principle} : dW(k,z) > 0 \rightarrow dQ(k,z) < 0$$

Then, it is immediate to formulate a synthetic version of both preceding formulas:

$$dQ(k,z) = -A \cdot dW(k,z), \text{ where } A > 0$$

This is the fundamental law of **Cl-1**. It is easy to see that, as is usual in such kinds of laws, this formula by itself is almost devoid of empirical content since $Q$ and $W$, taken in isolation, cannot be concretely determined (they are **Cl-1**-theoretical). They have to be put in some kind of testable relationship with the other (non-theoretical) primitive magnitudes of the theory: $P, V, t$. And, in fact, in his argumentation, Clausius always presupposes that $Q$ and $W$ are *dependent* on these magnitudes at least. Speaking in more formal terms, it may be said that $Q$ and $W$ are *functionals* of $P, V, t$, that is, there are functions $f^Q$ and $f^W$ (which are functions of functions) expressing the general form of the dependence of $Q$ and $W$ on $P, V, t$.

The preceding considerations allow now for a formal definition of the models of the basic theory-element of the net **Cl-1**. To abbreviate, we may skip out the explicit indication of the formal features of each single component of the models (i.e. whether it is a finite or infinite set, a differentiable function, and so on).

**Cl-1** : $x \in \mathbf{M}_0[\mathbf{Cl\text{-}1}]$ **iff** :$\exists K, Z, P, V, t, Q, W, A, f^Q, f^W$

such that

(0) $x = \langle K, Z, \Re, A, P, V, t, Q, W \rangle$

(1) $\forall k \in K \; \forall z \in Z : Q(k, z) = f^Q(P(k, z), V(k, z), t(k, z), \ldots)$

(2) $\forall k \in K \; \forall z \in Z : W(k, z) = f^W(P(k, z), V(k, z), t(k, z), \ldots)$

(3) $A > 0$

(4) $\forall k \in K \; \forall z \in Z : dQ(k, z) = -A \cdot dW(k.z)$

We could synthesize conditions (1)–(4) in one 'big' formula:

$$(\mathbf{FL_{Cl\text{-}1}}) \; \forall k \in K \; \forall z \in Z : df^Q(P(k, z), V(k, z), t(k, z), \ldots)$$
$$= -A \, df^W(P(k, z), V(k, z), t(k, z), \ldots), \text{ for } A > 0.$$

This is the theory's fundamental law.

What about the intended applications of this basic theory-element? Clausius himself clarifies this point as he writes: '[in this work] we are going to restrict our attention to permanent [i.e. ideal] gases and to steam at a maximum of its density' (Clausius, 1850: 11). Therefore, the basic domain of intended applications of **Cl-1**, **I[Cl-1]** is

$$\mathbf{I[Cl\text{-}1]} = \{\text{ideal gases; steam at a maximum density}\}.$$

Thus, the net's basic theory-element is: $\langle \mathbf{M}_0[\mathbf{Cl\text{-}1}]; \mathbf{I[Cl\text{-}1]} \rangle$, as previously defined.

## 8.5 The specializations of Cl-1

Two main specialization 'branches' may be identified in the tree representing the theory-net of **Cl-1**: the one corresponding to the study of ideal gases and the one aimed at steam at a maximum density. The corresponding specializations are obtained by specifying, in different ways, the functionals $f^Q$ and $f^W$ of the fundamental law.

For the *ideal gases*, Clausius takes up some previous results to propose the following specifications:

$$f^W = \frac{R\, dV\, dt}{V} \quad \text{(p. 15)}$$

$$f^Q = \left( \frac{d}{dt}\left(\frac{dQ}{dV}\right) - \frac{d}{dV}\left(\frac{dQ}{dt}\right) dV\, dt \right) \quad \text{(p. 17)}$$

As Clausius himself recognizes, the second specification is somewhat problematic since the notion of a 'heat differential' is itself physically and mathematically dubious.

However, by combining it with the less problematic determination of $f^W$, Clausius comes to the conclusion that one can establish a more plausible differential equation if we *postulate* that there must exist an 'arbitrary' function $U$ that allows for a mathematically correct formulation. (It has already been remarked that this '$U$' is 'our' internal energy.)

$$dQ = dU + A \cdot \frac{a+t}{V} \cdot dV$$

This is the most general specialization for ideal gases. Using the additional primitive concepts we have already indicated above ($U$, $R$, $c_V$, $c_P$, $a$, and the implicit notion of a gaseous substance which is assigned to each concrete body by a function $\gamma$) we come to the following definition of the models of this specialization:

**G** : $x \in \mathbf{M}[\mathbf{G}]$iff : $\exists K, Z, \Gamma, \gamma, P, V, t, Q, W, U, R, c_V, c_P, A, a, f^Q, f^W$

such that

(0) $x = \langle K, Z, \Gamma, \gamma, \mathfrak{R}, A, a, P, V, t, Q, W, U, R, c_V, c_P \rangle \in$
(1) $x = \langle K, Z, \mathfrak{R}, A, P, V, t, Q, W \rangle \in \mathbf{M}_0[\mathbf{Cl\text{-}1}]$
(2) $\forall k \in K\; \forall z \in Z : dQ(k, z)$
$\qquad = dU(k, z) + A \cdot R(\gamma(k)) \cdot \dfrac{a + t(k, z)}{V(k, z)} \cdot dV(k, z)$

For Clausius, the second term of the sum in (2) represents the 'external work' done by the gas. One obtains a specialization of this specialization by just assuming that this work has simply the form '$p\, dV$'. In such a case, by a simple calculation we obtain the law of ideal gases. We may call this specialization of the specialization **G** 'IG':

**IG** : $x \in \mathbf{M}[\mathbf{IG}]$ iff : $\exists K, Z, \Gamma, \gamma, P, V, t, Q, W, U, A, a, f^Q, f^W$

such that

(0) $x = \langle K, Z, \Gamma, \gamma, \text{IR}, A, a, P, V, t, Q, W, U \rangle \in \mathbf{M}[\mathbf{G}]$

(1) $\forall k \in K \, \forall z \in Z : A \cdot R(\gamma(k)) \cdot \dfrac{a + t(k,z)}{V(k,z)} \cdot dV(k,z) = p(k,z) \, dV(k,z)$

A further specialization on this same line consists in what Clausius calls an 'auxiliary hypothesis' (Clausius, 1850: 24) about the *molecular* constitution of ideal gases: Such gases, when they receive heat, do not do an 'internal work'; consequently, all the work done is 'exterior' and $U$ is solely a function of temperature. By abbreviating a little, we may define this specialization as follows:

**Mol** : $x \in \mathbf{M}[\mathbf{IG}] \wedge \exists \phi \, \forall k \in K \, \forall z \in Z \, (U(k,z) = \phi(t(k,z))$

Subsequently, Clausius makes an additional assumption: 'probably', in this case $\phi$ is a simple function determined only by $c_V$, which allows, in turn, for a determination of $c_P$:

**Mol** : $x \in \mathbf{M}[\mathbf{Mol}] \, \wedge \, U(k,z) = c_V(\gamma).t(k;z) \wedge c_P(\gamma(k)) - c_V(\gamma(k))$
$$= A \cdot R(\gamma)$$

In the models of **M[Mol]** one obtains as a corollary Poisson's law:

$$\forall \gamma \in \Gamma: c_P(\gamma)/c_V(\gamma) \text{ is a constant.}$$

The other branch of specializations in the net corresponds to *steam at a maximum density*. The specifications of the heat and work functionals are quite different from those of the ideal gases. And the additional primitive concepts needed (as has been indicated above) are also different. (Analogously to the case of ideal gases, for formal reasons, we need an assignment function $\pi$ assigning to each body considered a particular mixture of a liquid and a vaporous phase in the set $\Pi$ of all such kinds of mixtures.)

**V** : $x \in \mathbf{M}[\mathbf{V}]$ iff : $\exists K, Z, \Pi, P, V, t, Q, W, s, \sigma, m, c, h, A$

such that

(0) $x = \langle K, Z, \Pi, \pi, \Re, A, P, V, t, Q, W, s, \sigma, m, c, h \rangle \in \mathbf{Mo}[\mathbf{Cl\text{-}1}]$

(1) $\forall k \in K \; \forall z \in Z : dQ(k,z)$
$= df^Q(P(k,z), V(k,z), t(k,z), \ldots)$
$= \dfrac{dV(k,z)}{dt(k,z)} + c(\pi(k)) - h(k,z) \cdot dm(k,z) \cdot dt(k,z) \wedge dW(k,z)$
$= df^W(P(k,z), V(k,z), t(k,z), \ldots)$
$= (s(k) - \sigma(k)) \cdot dP(k,z) \cdot dm(k,z)$

Subsequently, Clausius introduces an additional hypothesis suggested by the experimental results of Regnault and Pambour, namely that $h$ always is negative.

$$V : x \in \mathbf{M}[V] \wedge \forall k \in K \; \forall z \in Z : h(k,z) < 0$$

We are now in a position to build up the graph representing the net **Cl-1** where the arrows represent successive specializations (see Fig. 8.2).

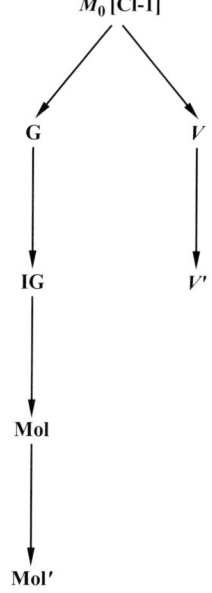

Figure 8.2

## 8.6 The basic theory-element of Clausius-2.1

In his 1854 paper, Clausius makes a new attempt at establishing phenomenological thermodynamics on a well-founded basis. This attempt is more restricted and less concrete than **Cl-1**: It almost exclusively deals with *cycles* ('*Kreisprozesse*'). He defines a cycle as a 'transformation' ('*Verwandlung*') whereby a body comes back to its original state. Another important difference with respect to the first article is that Clausius presents now the 'two principles of thermodynamics' in a completely independent manner, as if they would correspond to two different theories. The article is divided in two sharply separated parts with quite different titles, the second part being much longer and better articulated. Even by the style of exposition, one gets the impression that the two parts were written at two different periods of the development of Clausius's thought. The two principles now get the wording:

(I) 'Principle of the equivalence of heat and work'
(II) 'Principle of the equivalence of transformations'.

Formally, there is no other way but to reconstruct the two parts as two different theory-nets, which we will call '**Cl-2.1**' and '**Cl-2.2**', respectively. Let us analyse first **Cl-2.1**. Other than in the 1850 article, Clausius now states the heat/work equivalence principle by immediately introducing the 'anonymous' function $U$ as a basic concept, and he makes the following symptomatic remark about it: '$U$ [is a magnitude] that we cannot specify at present, but of which we know at least that it is completely determined by the initial and the final state of the body' (Clausius, 1854: 130). Consequently, he formulates the fundamental law of **Cl-2.1** as follows:

$$Q = U + A \cdot W$$

*Nota bene* : The symbol '$W$' employed here by Clausius does not now refer to the 'total work', as in **Cl-1**, but only to the 'external work'.

The first specialization considered refers to cycles. For such cases, Clausius writes: $U = 0$. Presumably, what he wants to express is what we now would write as

$$\oint dU = 0$$

A second specialization considered in this part of the paper is devoted to those bodies where the pressure is the same on all points (the intended applications being here gases, liquids and some solids). For this, Clausius proposes the equation

$$dQ = dU + A \cdot p \cdot dV$$

Towards the end of this first part of the essay, Clausius suggests that one could consider further specializations of this specialization 'by applying this formula to certain kinds of bodies. For the two most important cases, namely for ideal gases and for steam at a maximum density, I have developed this more special application in my precedent essay' (Clausius, 1854: 133). Obviously, he is referring to the net **Cl-1**.

Let us undertake now the structuralist reconstruction of **Cl-2.1**. Clearly, the conceptual framework is the same as for the models of the basic theory-element of **Cl-1** but for the additional primitive concept '$U$'. That is, the basic models have the form $\langle K, Z, \Re, P, V, t, Q, W, U\rangle$. Formally, this means that the models of $\mathbf{M}_0[\mathbf{Cl\text{-}1}]$ are now substructures of the models of $\mathbf{M}_0[\mathbf{Cl\text{-}2.1}]$. $\mathbf{M}_0[\mathbf{Cl\text{-}2.1}]$ is determined by the same conditions as $\mathbf{M}_0[\mathbf{Cl\text{-}1}]$, except for the introduction of an additional 'functional' to express the dependence of $U$ with respect to $V$ and $t$ (p. 131) and for the new form of the fundamental law:

**Cl-2.1** : $x \in \mathbf{M}_0[\mathbf{Cl\text{-}2.1}]$ iff : $\exists K, Z, P, V, t, Q, W, A, U, f^Q, f^W, f^U$

such that

(0) $\chi = \langle K, Z, \Re, A, P, V, t, Q, W, U\rangle$

(1) $\forall k \in K \, \forall z \in Z : Q(k, z) = f^Q(P(k, z), V(k, z), t(k, z))$

(2) $\forall k \in K \, \forall z \in Z : W(k, z) = f^W(P(k, z), V(k, z), t(k, z))$

(3) $A > 0$

(4) $\forall k \in K \, \forall z \in Z : U(k, z) = f^U(V(k, z), t(k, z)$

(5) $\forall k \in K \, \forall z \in Z : Q(k, z) = U(k, z) + A \cdot W(k, z)$

### 8.7 Specializations of Cl-2.1

The first specialization of this theory-net deals with cycles. Intuitively, a cycle is a process, that is, in our terms, a sequence $Z$ of states with a first element (the 'initial state') and a last element (the 'final state') such that the values of all fundamental magnitudes are the same in the two states. We could define this notion in completely formal terms within our model-theoretic framework, but since this involves some technicalities, for reasons of brevity, let us just presuppose that the notion of a cycle has already been defined. Then, the specialization looks like this:

**C** : $x \in \mathbf{M}[\mathbf{C}]$ iff : $\exists K, Z, P, V, t, Q, W, A, U$

such that

(0) $x = \langle K, Z, \Re, A, P, V, t, Q, W, U\rangle$ and $x \in \mathbf{M}_0[\mathbf{Cl\text{-}2.1}]$

(1) $Z$ is a cycle

(2) $\forall k \in K \, \forall z \in Z: \oint dU(k, z) = 0$

As remarked, the other specialization Clausius considers in this part of the article is completely independent of the previous one: In it, it is assumed that $W$ takes a particularly simple form: $P\,dV$. We may call it the 'uniform pressure specialization', UP.

**UP** : $x \in \mathbf{M}[\mathbf{UP}]$ iff : $\exists K, Z, P, V, t, Q, W, A, U$

such that

(0) $x = \langle K, Z, \Re, A, P, V, t, Q, W, U \rangle$ and $x \in \mathbf{M}_0[\text{Cl-2.1}]$

(1) $\forall k \in K \; \forall z \in Z: dQ(k,z) = dU(k,z) + A \cdot P(k,z) \cdot dV(k,z)$

As Clausius himself indicates, from this latter specialization one could obtain the law of ideal gases as well as the laws for steam systems. This means that we can *embed* one part of the previous theory-net **Cl-1** into the new net **Cl-2.1**. This is a typical move in a crystallization process.

The graph of the theory-net **Cl-2.1** is accordingly (see Fig. 8.3).

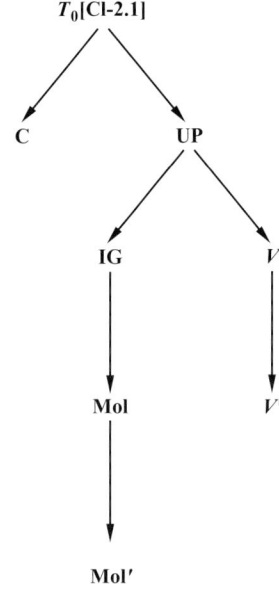

Figure 8.3

### 8.8 The basic theory-element of Clausius-2.2

As already noted, the theory-net **Cl-2.2** is exclusively devoted to the most general treatment of the 'second principle' as completely independent of the 'first principle'. And its intended domain of applications are solely cycles.

To begin with, Clausius provides an intuitive formulation of the 'second principle' that he considers to be more general than the one attributed to previous researchers (notably Carnot and Clapeyron): 'Heat can never go from a cold to a hotter body unless another transformation takes place that depends on the first one' (Clausius, 1854: 134). Clausius sets himself the task of finding a mathematically appropriate formulation of this principle. The first step consists in introducing the new notion of 'equivalence value' ('*Aequivalenzwerth*') of a transformation (of heat into work or conversely). His proposal is

$$\text{Equivalence value of } W: Qf(t) =_{df} Q \cdot \frac{1}{T(t)}$$

Here there is a new 'anonymous' function $T$, of which Clausius says: '$T$ is an unknown function of the temperature appearing in the equivalence values' (Clausius, 1854: 143). This is for us again an example of a $T$-theoretical magnitude. Only at the end of the paper, Clausius *suggests* that $T$ may simply be regarded as the *absolute temperature*.

Before the new formulation of the 'second principle', Clausius establishes a distinction between 'reversible' ('*umkehrbar*') and 'irreversible' ('*unumkehrbar*') cycles. The text does not contain any precise definition of what these terms mean; apparently, the reader is supposed to have a clear intuition of these notions. Anyway, after a long and cumbersome argument, Clausius comes to the conclusion that the formal condition for reversible cycles is

$$\oint \frac{dQ}{T} = 0$$

A specialization of this condition is Clapeyron's equation

$$\frac{dT}{dt} \bigg/ T = \frac{A}{C}$$

where $C$ is the so-called 'Carnot function'.

For irreversible cycles, Clausius concludes that the condition should be

$$\oint \frac{dQ}{T} > 0.$$

If we admit the 'additional hypothesis' ('*Nebenannahme*'), that Clausius takes from Regnault, according to which 'an ideal gas, when expanding at constant temperature, absorbs only the quantity of heat needed for the external work done' (Clausius, 1854: 153), we obtain:

$$\frac{dQ}{dV} = A \cdot P$$

and if we add the ideal gas law, we get as a *theorem*

$$T = a + t, \text{ where } a \text{ presumably is } 273$$

That is why we can consider $T$ to be the absolute temperature.

Clausius never explicitly formulates the fundamental law common to both reversible and irreversible cycles but it is clear that this can be nothing but

$$\oint \frac{dQ}{T} \geq 0$$

Let us proceed now to the structuralist representation of the various theory-elements of **Cl-2.2**. First, it is to be noted that its conceptual framework is a conceptual extension of **Cl-1** since the models have to contain now a new magnitude, $T$. Formally, this means that the models of **Cl-1** reappear now as *substructures* of the models of **Cl-2.2**, which have the form $\langle K, Z, \Re, P, V, t, Q, W, T\rangle$. These models are determined by the fundamental law expressing the relationship between heat and absolute temperature in all kinds of cycles.

**Cl-2.2** : $x \in \mathbf{M}_0[\mathbf{Cl\text{-}2.2}]$ iff : $\exists K, Z, P, V, t, Q, W, A, T, f^Q, f^W, f^T$ such that

(0) $x = \langle K, Z, \Re, A, P, V, t, Q, W, T\rangle$

(1) $Z$ is a cycle

(2) $\forall k \in K \; \forall z \in Z : Q(k, z) = f^Q(P(k, z), V(k, z), t(k, z))$

(3) $\forall k \in K \; \forall z \in Z : W(k, z) = f^W(P(k, z), V(k, z), t(k, z))$

(4) $A > 0$

(5) $\forall k \in K \; \forall z \in Z : T(k, z) = f^T(t(k, z))$

(6) $\forall k \in K \; \oint \dfrac{dQ(k, z)}{T(k, z)} \cdot dz \geq 0$

## 8.9 Specializations

A first specialization concerns reversible cycles.

**REV** : $\mathbf{x} \in \mathbf{M}[\mathbf{REV}]$ iff :

(0) $x \in \mathbf{M}_0[\mathbf{Cl\text{-}2.2}]$

(1) $\forall k \in K \; \oint \dfrac{dQ(k, z)}{T(k, z)} \cdot dz = 0$

This specialization may, in turn, be specialized in order to obtain Clapeyron's law. For this, a new $T$-theoretical magnitude is introduced: 'Carnot's function', $C$. In this way, the elements of **M[REV]** become substructures of the Clapeyron models.

**CLAP** : $\mathbf{x} \in \mathbf{M}[\mathbf{CLAP}]$ iff :

(0) $x = \langle K, Z, \Re, A, P, V, t, Q, W, T, C\rangle$

(1) $\langle K, Z, \mathrm{IR}, A, P, V, t, Q, W, T\rangle \in \mathbf{M}[\mathbf{REV}]$

(2) $\forall k \in K \; \forall z \in Z : \dfrac{1}{T(k, z)} \cdot \dfrac{dT(k, z)}{dt(k, z)} = \dfrac{A}{C(k, z)}$

The other specialization line concerns irreversible cycles.

**IRREV** : $x \in$ **M[IRREV]** iff :

(0) $x \in$ **M[Cl-2.2]**

(1) $\forall k \in K \oint \dfrac{dQ(k,z)}{T(k,z)} \cdot dz > 0$

From this, we may obtain a further specialization by admitting Regnault's additional hypothesis above-mentioned.

**REGN** : $x \in$ **M[REGN]** iff :

(0) $x \in$ **M[IRREV]**

(1) $\forall k \in K \; \forall z \in Z : \dfrac{dQ(k,z)}{dV(k,z)} = A \cdot P(k,z)$

Within this same branch of the net, we may get still another specialization by simply incorporating the ideal gas law we have already reconstructed, **M[IG]**. It is not necessary to repeat its formulation here.

By combining **M[REGN]** with **M[IG]**, that is, by constructing a specialization common to both specializations that we may call '**AT**' (for 'absolute temperature'), or in model-theoretic terms, by taking **M[AT]** = **M[REGN]** ∩ **M[IG]**, the following result may be obtained:

If $x \in$ **M[AT]**, then $\forall k \in K \; \forall z \in Z : T(k,z) = 273 + t(k,z)$

Accordingly, the theory-net **Cl-2.2** has the form shown in Fig. 8.4:

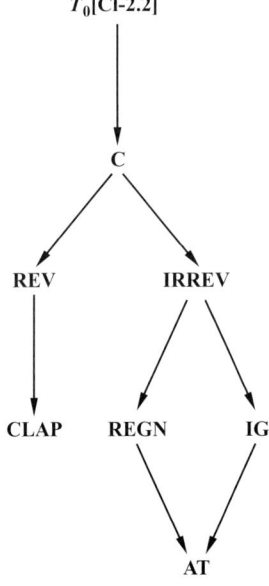

Figure 8.4

We may now compare the three theory-nets implicit in Clausius's writings examined in this chapter, **Cl-1**, **Cl-2.1** and **Cl-2.2**. It is to be noted that some (but not all) specializations of **Cl-1** are *embedded* into the nets **Cl-2.1** and **Cl-2.2**, while some of **Cl-2.1** are embedded into **Cl-2.2**.

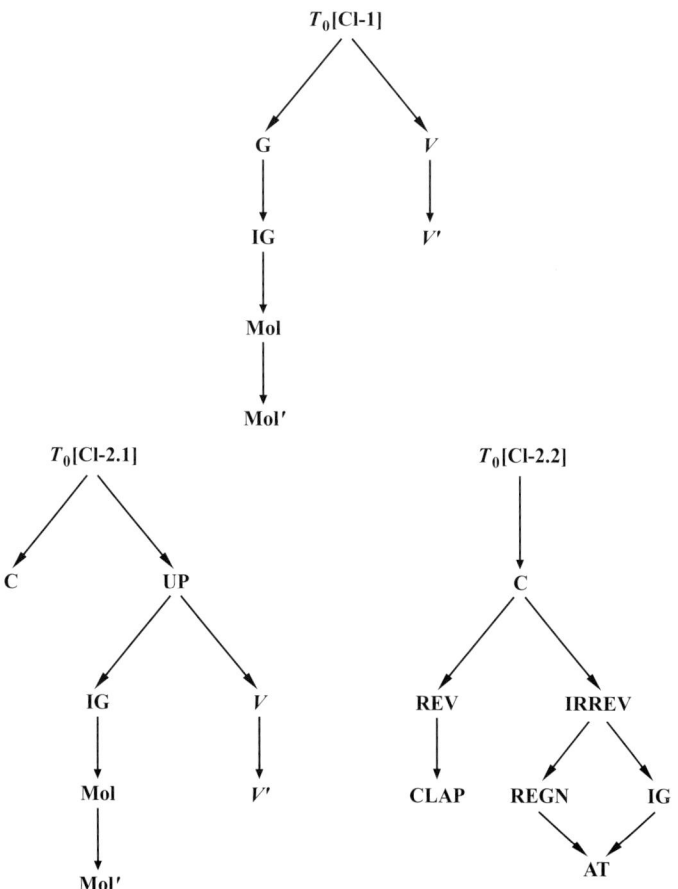

Figure 8.5

## References

Balzer, W., Moulines, C. U. and Sneed, J. D. (1987). *An Architectonic for Science*. Dordrecht: Reidel.

Caamaño, M. C. (2004). El problema de la inconmensurabilidad de las teorías científicas. Ph.D. thesis, University of Santiago de Compostela.

Callen, H. B. (1960). *Thermodynamics*. New York: Wiley.

Clausius, R. J. (1850). Über die bewegende Kraft der Wärme. *Annalen der Physik*, **79**, 368–397, 500–524.

Clausius, R. J. (1854). Über eine veränderte Form des zweiten Hauptsatzes der mechanischen Wärmetheorie. *Annalen der Physik*, **93**, 481–506.

Clausius, R. J. (1857). Über die Art der Bewegung, die wir Wärme nennen. *Annalen der Physik*, **100**, 497–507.

Elkana, Y. (1974). *The Discovery of the Conservation of Energy*. London: Hutchinson.

Gibbs, J. W. (1876). On the equilibrium of heterogeneous substances, Part I. *Transactions of the Connecticut Academy of Arts and Sciences* 3. Reprinted in: *The Scientific Papers of J. Willard Gibbs*, ed. H. A. Bumstead and R. G. van Name. New York: Longmans, Green & Co, 1906 (2nd edn, New Haven CT: Yale Univ. Press, 1957).

Moulines, C. U. (1991). The classical spirit in J. Willard Gibbs's classical thermodynamics. In *Thermodynamics: History and Philosophy*, ed. K. Martinás, et al. Singapore: World Scientific.

Moulines, C. U. (1996). Zur Typologie wissenschaftlicher Entwicklung nach strukturalistischer Deutung. In *Cognitio humana – Dynamik des Wissens und der Werte*, ed. C. Hubig. Leipzig: Akademie-Verlag, pp. 397–410.

Sneed, J. D. (1971). *The Logical Structure of Mathematical Physics*. Dordrecht: Reidel (2nd revised edition 1978).

Stegmüller, W. (1973). *Theorienstrukturen und Theoriendynamik*. Berlin: Springer-Verlag.

# 9
# Reduction and renormalization

ROBERT W. BATTERMAN

## 9.1 Introduction

In this chapter I want to consider the so-called reduction of thermodynamics to statistical mechanics from both historical and relatively contemporary points of view. As is well known, most philosophers not working in the foundations of statistical physics still take this reduction to be a paradigm instance of that type of intertheoretic relation. However, numerous careful investigations by many philosophers of physics and physicists with philosophical tendencies show this view is by and large mistaken. It is almost surely the case that thermodynamics does not reduce to statistical mechanics according to the received view of the nature of reduction in the philosophical literature. What is interesting is that, while not framing the issue in these terms, J. Willard Gibbs can also be seen as being somewhat sceptical about the possibility of a philosophical reduction of thermodynamics to statistical mechanics. Gibbs' scepticism is, of course well-known. Nevertheless, I think his remarks bear further consideration given certain advances in understanding the foundations of statistical physics.

I will first briefly run over some philosophical ground, outlining the received approach to theory reduction as well as what I take to be a more promising conception of reduction that parallels, to some extent the way physicists typically speak of theory reduction. Following this I will discuss Gibbs' famous caution in connecting thermodynamical concepts with those from statistical mechanics. This is presented in chapter XIV, 'Discussion of Thermodynamic Analogies', of his book *Elementary Principles in Statistical Mechanics*. We will see that there are several reasons Gibbs held back from identifying thermodynamic quantities such as temperature and entropy with specific statistical mechanical quantities. I will then present a sketch of a program for reduction that involves deep connections between results in probability theory on limit theorems and the so-called real space renormalization techniques that play such an essential role in understanding the universality of critical phenomena. The framework provides a way of unifying two fundamental problems – the so-called equivalence of ensembles and the existence of critical phases.

## 9.2 Philosophical reduction

Most contemporary views about reduction owe much to the seminal work of Ernest Nagel. In *The Structure of Science* Nagel (1961:338) asserts that 'reduction... is the explanation of a theory or a set of experimental laws established in one area of inquiry, by a theory usually though not invariably formulated for some other domain'. Standard views about explanation hold that to explain some phenomenon requires the derivation of a statement characterizing that phenomenon from laws of some true theory. Thus, reduction, very roughly, involves the derivation of the laws of the 'reduced' theory from the laws of the 'reducing' theory. In the case of thermodynamics we have a reduction to statistical mechanics because the laws of thermodynamics are (supposedly) derivable from those of statistical mechanics.

Of course, Nagel himself realized that to actually carry out such derivations requires non-trivial work. Here is one major problem: In most cases the laws of the reduced theory (thermodynamics) contain terms that do not appear in the laws of the reducing theory (statistical mechanics). For example, while thermodynamics talks of temperature and entropy, statistical mechanics does not. Some kind of connection between the predicates appearing in the reduced theory and those in the reducing theory (provided by so-called 'bridge laws') is required so that a *derivation* of the laws of the one from the laws of the other will be possible. These bridge principles are essential, because explanation and reduction are conceived to be arguments in a formal language. They are linguistic relations that obtain between the theories understood as sets of sentences in that formal language.

As I mentioned, it is commonplace to read that bridge laws or connections can be found in the thermodynamics/statistical mechanics case. For instance, one often sees that 'temperature' in thermodynamics is to be identified with 'mean molecular kinetic energy'. This is taken to be a paradigm example of the sort of bridge law Nagel demands for reduction.

There are many reasons to be sceptical that a Nagelian or neoNagelian reduction of thermodynamics to statistical mechanics is possible. One worry concerns the status of the bridge laws. In one sense they seem to be statements of definition. But surely they cannot be logical connections – statements true solely in virtue of the meanings of the terms and, perhaps, knowable merely by reflection upon those meanings. Another possibility is that the bridge principles are stipulated conventions or coordinative definitions. A third possibility is that they express factual claims that essentially have the status of physical hypotheses. (Nagel, 1961: 354)

In the context of the reduction of thermodynamics these worries become acute. The 'temperature equals mean molecular kinetic energy' bridge law *identifies* a fundamentally *non-statistical* quantity with a fundamentally statistical quantity. How is this supposed to work? Of course, this problem will arise as well for other strictly thermodynamic predicates such as entropy.

Actually, the situation is even more dire. Even terms such as 'pressure' require appropriate bridge laws. Surely it is correct to associate in some way the thermodynamic pressure with some function of the number of collisions per unit area upon the walls of the container. But which function? As Sklar points out,

there is, for a particular sample of gas at equilibrium, the actual momentum transferred by the molecules..., and there is its average value per unit area per unit... time. On the other hand, there is the quantity calculated for an ensemble of similarly constituted systems..., or, alternatively by looking for the most probable value of the relevant quantity in the ensemble. Whereas the former sort of pressure, the feature of the individual system, will be expected to fluctuate, the latter kinds of ensemble quantities..., will, of course, not. Here fluctuations will show up as assimilated into the ensemble description by the calculation of averages or most probable values for quantities, but the averages themselves are not the sort of things to fluctuate.

*(Sklar, 1993: 349–350)*

From the point of view of these concerns about the philosophical understanding of theory reduction, it is natural to interpret Gibbs' famous chapter XIV, 'Discussion of Thermodynamic Analogies' to be an investigation into the nature and status of the bridge laws required for the reduction of thermodynamics to statistical mechanics. Sklar, following the above discussion of the difficulties encountered in trying to identify thermodynamic concepts with statistical mechanical concepts, says the following:

It should not surprise us that Gibbs, when he came to associate ensemble quantities with thermodynamic quantities in Chapter XIV of his book, spoke of the "thermodynamic analogies" when he outlined how thermodynamic functional interrelations among quantities were reflected in structurally similar functional relations among ensemble quantities. He carefully avoided making any direct claim to have found what the thermodynamic quantities "were" at the molecular dynamical level.

*(Sklar, 1993: 350)*

Gibbs was surely cautious and he did care about the connections between thermodynamics and statistical mechanics. But I think that this may not be the most fruitful way to think about his caution or his approach to intertheoretic relations.

## 9.3 Gibbs' caution and the 'non-equivalence' of ensembles

In the Preface to *Elementary Principles in Statistical Mechanics* Gibbs (1981: vi–vii) asserts that the laws of statistical mechanics for conservative systems of finite degrees of freedom are exact. On the other hand, the laws of thermodynamics as

> empirically determined, express the approximate and probable behavior of systems of a great number of particles, or more precisely, they express the laws of mechanics for such systems as they appear to beings who have not the fineness of perception to enable them to appreciate quantities of the order of magnitude of those which relate to single particles, and who cannot repeat their experiments often enough to obtain any but the most probable results.
>
> *(Gibbs, 1981: vi)*

Further, the fact that these laws for finite systems are exact,

> ... does not make them more difficult to establish than the approximate laws for systems of a great many degrees of freedom, or for limited classes of such systems. The reverse is rather the case, for our attention is not diverted from what is essential by the peculiarities of the system considered, *and we are not obliged to satisfy ourselves that the effect of the quantities and circumstances neglected will be negligible in the result.*
>
> *(Gibbs, 1981: vii; my emphasis)*

This last clause suggests that Gibbs recognizes that showing how the 'approximate' laws – by which he means the laws of thermodynamics – will emerge out of the 'exact' formulas for systems with finite and relatively small numbers of degrees of freedom will be difficult. In addition, this passage suggests that something is to be gained by not being distracted by specific details or peculiarities of the systems under investigation.

I think this latter suggestion is completely correct, although odd in the circumstances. It suggests that the statistical approach is to be preferred to that of rational thermodynamics, because we will focus on statistical properties rather than upon specific features of individual systems. The odd part, is that for Gibbs the statistical mechanical formulas for finite systems are exact, and so, presumably will provide all of the details of the actual system of interest. Thus, from Gibbs' point of view *exact* expressions are those statistical mechanical expressions involving the actual finite sums corresponding to the actual finite number of particles in the system under investigation. From a more contemporary point of view, the notion of an 'exact' expression has a different meaning: In that context one obtains exact or precise results only by taking the thermodynamic limit where our inability to discern the precise number of particles becomes irrelevant, and *where sharp (i.e. exact), as opposed to fluctuating, values emerge.*

Forty odd years later, Khinchin, in the context of discussing Gibbs' priority in the 'systematic exposition of the foundations of statistical mechanics', focuses on the statistical point of view. He says that

It was precisely the necessity of a statistical foundation for the general laws of thermodynamics that produced trends which found their expression in the construction of statistical mechanics. To avoid making any special hypotheses about the nature of the particles it became necessary in establishing a statistical foundation to develop laws which had to be valid no matter what was the nature of those particles (within quite wide limitations).
*(Khinchin, 1949: 3)*

I think that Khinchin here is providing a more modern spin to Gibbs' discussion than is actually present in the discussion itself. Despite this, and as Khinchin stresses, I think that Gibbs was one of the first to appreciate the idea that sometimes, in the investigation and understanding of general principles or laws, too much attention to the details can get in the way. Comparatively speaking, we are able to easily formulate the laws of statistical mechanics because our 'attention is not diverted from what is essential'.

Gibbs discusses a number of reasons why it is important to avoid making assumptions or hypotheses about the nature of the particles or systems under investigation. These reasons, I believe, can fruitfully be partitioned into two classes: pragmatic reasons and theoretical/fundamental reasons. Let me discuss these in turn.

Gibbs was working at a time in which physics was beginning a period of what Kuhn called 'extraordinary' science. The classical paradigm was beginning to succumb to a number of anomalies whose ultimate resolution required the quantum theory. Gibbs was well aware of these anomalies and they provided pragmatic reasons for him to be sceptical of detailed hypotheses about the nature of the systems. For example, Gibbs refers to the problem of a gas composed of diatomic molecules for which according to the equipartition theorem there should be six degrees of freedom sharing the energy, whereas experiments on the specific heat of such a gas indicate that only five degrees of freedom apparently share in the energy. The other example he mentions in several places concerns the phenomenon of radiant heat. Here the problem is that the assumption that we are dealing with systems of finite degrees of freedom, does not appear to be adequate for the explanation 'of the properties of bodies' (Gibbs, 1981: 167).

These experimental and theoretical anomalies clearly contributed to Gibbs' inclination not to formulate specific hypotheses about the nature of various systems. Surely, they were significant pragmatic factors relating to his caution toward providing 'true' as opposed to merely analogical connections between thermodynamic concepts and statistical mechanical concepts.

Nevertheless, as I noted, there is a more fundamental reason to eschew the making of specific hypotheses. Paradoxically, it is also a reason, at least in retrospect, for thinking that his caution and repeated disclaimers ultimately were *not* completely warranted. I will call this fundamental worry the problem of the 'non-equivalence of ensembles'.

## 9.4 The canonical ensemble

As is well known, Gibbs' canonical ensemble is appropriate for representing equilibrium systems that can be considered to be in thermal contact with a large heat bath at a fixed temperature. Here the idea is that each member of this ensemble remains at constant energy but that the members take on all possible values for energy. The demand that this ensemble represents statistical equilibrium is the demand that the probability density $\rho(q, p)$ be invariant over a region in phase space under its dynamical evolution. Any function of phase that is constant along a phase space trajectory (hence any function of the energy) satisfies this demand, but Gibbs argues that one such function, in particular, has special features worthy of representing thermodynamic equilibrium:

$$\rho = e^{\frac{\psi - \epsilon}{\Theta}} \tag{9.1}$$

Here $\epsilon$ is the energy of a system and $\psi$ and $\Theta$ are constants. Gibbs argues that one should take the negative of the 'index of probability'

$$-\log \rho = -\frac{\psi - \epsilon}{\Theta} \tag{9.2}$$

to be *analogous* to the thermodynamic entropy and the modulus, $\Theta$, to be *analogous* to the temperature. Furthermore, Gibbs argued that the deviations in energy from the mean value for energy in the ensemble should be small. In fact, he claims that those deviations should be 'of the same order of magnitude as the reciprocal of the number of degrees of freedom, and therefore to human observation the individual values are indistinguishable from the average values when the number of degrees of freedom is very great' (Gibbs, 1981: 168). In such a situation the deviations in the index of probability from its mean will also be negligible in this operational sense.

## 9.5 The micro-canonical ensemble

Gibbs also discusses the micro-canonical ensemble[1] which is appropriate for representing equilibrium systems that can be considered to be thermally isolated from

---

[1] Here we ignore the third important ensemble of Gibbs: The grand canonical ensemble which is appropriate for the description of systems that can exchange particles with the environment.

the rest of the world. Here we have a collection of systems that all have the same energy. In such an ensemble the quantity

$$\log V \tag{9.3}$$

is the *analogue* of the thermodynamic entropy where $V$ is the 'size' or 'volume' (in the natural measure) of the phase space region to which the possible micro-states of the systems are confined by macroscopic constraints. The *analogue* of temperature in the micro-canonical ensemble is the quantity

$$\frac{d\epsilon}{d\log V} \tag{9.4}$$

Gibbs says

we have thus precisely defined quantities, and rigorously demonstrated propositions, which hold for any number of degrees of freedom, and which, when the number of degrees of freedom ($n$) is enormously great, would appear to human faculties as the quantities and propositions of empirical thermodynamics.

(Gibbs, 1981: 169)

Nevertheless, it is well known that different quantities may approach the same form in some appropriate limit, say, $n \to \infty$.

There may be therefore, and there are, other quantities [other than the modulus, $\Theta$, and the index of probability, $\frac{\psi-\epsilon}{\Theta}$,] which may be thought to have some claim to be regarded as temperature and entropy with respect to systems of a finite number of degrees of freedom.

(Gibbs, 1981: 169)

Of course, one such set of quantities are those derived from the micro-canonical distribution: Equations (9.4) and (9.3), respectively.

Gibbs takes the fact that different, exact, statistical physical quantities apparently have equal claim to be the analogs of thermodynamic temperature and entropy to be problematic. This is the 'non-equivalence' of ensembles and it is the theoretical or fundamental reason to resist claims to have found the (statistical) mechanical definition of those quantities.

## 9.6 Limiting reductions/limiting relations between theories

So Gibbs realized that in the thermodynamic limit certain quantities that differ for finite systems will have the same limiting form. He considered this a difficulty and it may be one reason that he focused on the exact statistical mechanical quantities for finite $n$. On the other hand, he also repeatedly talks of the laws of thermodynamics as being the approximate expressions of the exact statistical mechanical laws when the number of degrees of freedom gets large. The idea that suggests itself here is that the limit of statistical mechanics, as the number of degrees of freedom goes to infinity, should yield the continuum thermodynamic theory. Physicists often express this connection as an instance of reduction, though not in the same way as Nagel and his followers. On the physicists' view statistical mechanics reduces to thermodynamics (in the appropriate limit). Schematically we can represent this conception of reduction as follows:

$$\lim_{\epsilon \to 0} T_f = T_c \tag{9.5}$$

In the present case we can (somewhat sloppily) let $\epsilon = 1/N$ where $N$ is the number of particles, and take $T_f$ to be statistical mechanics and $T_c$ to be thermodynamics.

Philosophers, as we have seen, typically talk of reduction going the other way – that is, thermodynamics reduces to statistical mechanics via deductive derivation of the laws of the former from those of the latter. On my view, these two conceptions of reduction are related to one another. In fact, I believe that those cases where a physicists' limiting reduction holds – that is, where equality obtains in schema (9.5) – are cases for which the philosophers' conception of derivational reduction will likely hold as well (Batterman, 1995; 2002b). This is because we can take the limiting relations as providing us with something like the bridge laws appropriate for Nagel-like reduction.[2]

But limiting reductions in which the equality of schema (9.5) hold are, in fact, rare. Schema (9.5) will fail when the limiting relation is singular. That is to say, that the limiting behaviour as $\epsilon \to 0$ is qualitatively different than the behaviour *at* the limit when $\epsilon = 0$. Singular limits do not mean that there are no interesting connections between the theories. Usually, just the opposite. But in such cases I have argued it is best to give up talk of 'reduction' altogether and to speak instead of 'intertheoretic relations'.

Now I do not believe that the historical evidence is sufficient to enable us to decide whether Gibbs held a philosophical conception of reduction as deductive derivation or whether he opted for the physicists' limiting sense of reduction. I doubt that he conceptualized the issues in this way, and there are places where, with hindsight, he seems to be talking of each. Nevertheless, I would like to investigate the fruitfulness of the approach to intertheoretic relations that focuses on the importance of limits and asymptotic connections between theories.

I will focus on two issues. The first concerns the equivalence or non-equivalence of ensembles. We have just seen that one reason for Gibbs' scepticism is the existence of finite distinct *exact* quantities that apparently have equal claim to be identified with the thermodynamic concepts of entropy and temperature. Can one

---

[2] However, these will typically be mathematical relations and not universal biconditionals of the sort Nagel and others typically talk about.

resolve this problem by thinking of intertheoretic relations as limiting relations between theories as opposed to regarding them as derivational Nagelian type relations? A second problem is, oddly, nowhere mentioned in *Elementary Principles*. This is the issue of the existence and explanation of critical phenomena and phase transitions. It is of crucial importance and Gibbs was surely aware of it. In fact, I want to argue that the two problems are intimately related to one another. They can both be treated using a similar strategy based upon some deep limit theorems from probability theory. Furthermore, I hope to show that the fact that these two problems are related is relevant to the possibility of realizing some kind of partial reductive relation between thermodynamics and statistical mechanics after all. More cautiously, perhaps, I intend to show that it is possible to assert something stronger than a mere Gibbsian analogy between certain concepts in the two theories.

## 9.7 The renormalization group, the central limit theorem, and the *equivalence* of ensembles

Gibbs was aware that in the thermodynamic limit there is a kind of equivalence of ensembles. After all he states that there 'may be more than one proposition relating to finite vales of *n*, which approach the same limiting form for $n = \infty$' (Gibbs, 1981: 169). As we have seen he takes this to be problematic. From a more modern perspective, however – one which recognizes explicitly the essential role played by the thermodynamic limit – this can very well be considered a virtue. The idea here is that the equivalence of ensembles (to the extent that it can actually be demonstrated) is evidence of a kind of universality. The thermodynamic phenomenology obtains regardless of the exact statistical mechanical details.

One can see that this is to be expected by considering briefly Khinchin's program for statistical mechanics where the central limit theorem of probability plays a crucial role. Khinchin focuses on specific kinds of functions defined on phase space – so-called 'sum functions' – having the form

$$S(n) = \sum_{i=1}^{n} S_i$$

He employs the central limit theorem to show that the dispersions of the *suitably normalized* sum functions will, in the thermodynamic limit ($n \to \infty$), be distributed according to the normal or Gaussian distribution. Thus we should expect that phase functions of the right form – those that presumably represent thermodynamic quantities – will be peaked around their mean values with root mean square deviations proportional to $n^{1/2}$.

But the real question is why the Gaussian distribution should play such a fundamental role. Why should it emerge as the limiting distribution for statistical systems in equilibrium, virtually regardless of the nature of the systems' molecular details and regardless of whether it is best represented using the canonical or the micro-canonical distributions of Gibbs? To answer this, I will sketch an argument from probability theory that employs the renormalization scheme developed by Kadanoff, Fisher and Wilson. The connections between renormalization in statistical mechanics and probability theory have been stressed in a series of articles in the 1970s by Sinai, Cassandro, and, most forcefully, by Jona-Lasinio (Bleher and Sinai, 1973; Jona-Lasinio, 1975; Cassandro and Jona-Lasinio, 1978; Sinai, 1978, 1982). My presentation here follows Sinai's argument in (Sinai, 1992: Lecture 15).

Suppose we have a sequence of random variables $S_i$ such as spins on a one dimensional lattice. Suppose further that the $S_i$'s have finite second moments and are independent and identically distributed with the mean values, $E(S_i) = 0$.[3] Consider a subsequence of the integers, $n_p = 2^p$, and define the random variable, $S_p$ – a sum function of the spins – as follows:

$$S_p = \frac{1}{2^{p/2}} \sum_{i=1}^{2^p} S_i \qquad (9.6)$$

It follows that

$$S_{p+1} = 1/\sqrt{2}\left(S'_p + S''_p\right) \qquad (9.7)$$

where

$$S'_p = \frac{1}{2^{p/2}} \sum_{i=1}^{2^p} S_i \quad \text{and} \quad S''_p = \frac{1}{2^{p/2}} \sum_{i=2^p+1}^{2^{p+1}} S_i$$

For example, let $p = 2$ then

$$S'_2 = \frac{1}{2} \sum_{i=1}^{4} S_i = \frac{1}{2}(S_1 + S_2 + S_3 + S_4) \qquad (9.8)$$

$$S''_2 = \frac{1}{2} \sum_{i=5}^{8} S_i = \frac{1}{2}(S_5 + S_6 + S_7 + S_8) \qquad (9.9)$$

From (9.6) we have

$$S_3 = \frac{1}{2^{3/2}}(S_1 + \cdots + S_8)$$

---

[3] $E(\cdot)$ represents the expectation. The assumption of independence will be relaxed later.

But we see from (9.8) that $(S_1 + S_2 + S_3 + S_4) = 2S_2'$ and from (9.9) that $(S_5 + S_6 + S_7 + S_8) = 2S_2''$. Hence,

$$S_{p+1} = S_3 = \frac{1}{\sqrt{2}}\left(S_2' + S_2''\right)$$

So, $S_p'$ and $S_p''$ are independent and identically distributed random variables. The probability distribution for the sum of two independent random variables $\xi_1$, $\xi_2$, (that is, for the random variable $\xi = \xi_1 + \xi_2$) is given by the convolution of their individual distribution functions. Thus if $F_1$ is the distribution function for $\xi_1$ and $F_2$ is the distribution function for $\xi_2$, then

$$F_{1,2}(x) = \int_{-\infty}^{\infty} F_1(x - u)\, dF_2(u) \tag{9.10}$$

is the distribution function for the random variable $\xi$. Now, the random variables $S_p'$ and $S_p''$ are not only independent but they are also identically distributed. Let $F_p$ be the distribution function for $S_p$ – Equation (9.6) which is this common distribution function. Then from (9.10) we have

$$F_{p+1} = \int_{-\infty}^{\infty} F_p(\sqrt{2}x - u)\, dF_p(u) \tag{9.11}$$

which is the distribution function for the random variable $S_{p+1}$ (recall (9.7)). It follows that for any $p$, the distribution function $F_p$ for the variable $S_p$ can be obtained by iterating this non-linear convolution operation starting from the distribution function $F_0$ for $S_0$ – the random variable for an individual spin. Thus, let

$$TF = \int_{-\infty}^{\infty} F(\sqrt{2}x - u)\, dF(u) \tag{9.12}$$

for any distribution function $F$ then $F_{p+1} = TF_p$ and $F_p = T^p F_0$. The operator $T$ is the analogue, for the current case, of the renormalization group operator.

It can be shown that the Gaussian distribution,

$$G(x) = \frac{1}{\sqrt{2\pi}} \int_{-\infty}^{x} e^{-\frac{u^2}{2}}\, du$$

is a fixed point of the transformation $T$. In other words, $TG = G$. The emergence and the ubiquity of the Gaussian distribution is related to the stability of this fixed point in the space of distributions and that stability can be investigated by examining the stability of the linearization, $L$, of the non-linear operator $T$ in the neighbourhood of the fixed point $G$.

Sinai shows that the linear operator $L$ at the point $G$ is given by

$$Lh = \frac{\sqrt{2}}{\sqrt{\pi}} \int_{-\infty}^{\infty} h\left(u + \frac{x}{\sqrt{2}}\right) e^{-\frac{u^2}{2}} du$$

One then determines the spectrum and eigenfunctions of $L$. It turns out that the eigenfunctions are Hermite polynomials $h = P_k(x)$ with eigenvalues

$$u_k = 2^{1-\frac{k}{2}}$$

for $k = 0, 1, 2, \ldots$. It follows that $u_0, u_1 > 1$, $u_2 = 1$, and for all $k > 2$, $u_k < 1$. Thus, the 'directions' $P_0$ and $P_1$ are unstable, $P_2$ is marginal and all other directions are stable. This analysis tells us that the fixed point $G$ is stable for linear approximation with respect to perturbations that lie in the *class* of distribution functions $F$ with zero expectation and finite variance (Sinai, 1992: 132). And so, a large class of distribution functions will behave just like the Gaussian in the limit $n_p \to \infty$. This class of distributions, in the jargon of renormalization group theory, is a 'universality class'.

The upshot of this analysis for the problem of the equivalence of ensembles is that we must expect those finite systems distributed according to the microcanonical distribution, and those distributed according to the canonical distribution, to behave in the same way as the number of their components gets large. In fact, we can use the Gaussian distribution to calculate the dispersions of various quantities from their means for large finite systems of either type. A further consequence of this form of argument is that in order to solve certain problems regarding universal behaviour, one need only work with the most convenient 'system' in the relevant universality class. Nigel Goldenfeld calls such a system a 'minimal model' and asserts that it is a model that 'most economically caricatures the essential physics' (Goldenfeld 1992: 33). In the current context, Gibbs' insistence that for most problems it is easier to calculate using the canonical distribution, is an indication that the canonical distribution may be considered to be a minimal model.[4]

While Gibbs took the existence of the different quantities associated with the different ensembles – each apparently having equal claim to be analogues of thermodynamic temperature and entropy – to be a problem, from the current perspective, we should take their existence to be a virtue: In the thermodynamic limit, we can demonstrate the *equivalence* of the ensembles, and we can show that calculations with either ensemble will yield the thermodynamic phenomenology that we are after. On this way of thinking, the question of what quantity is to be

---

[4] See (Batterman, 2002a) for a discussion of minimal models and asymptotics.

offered as *the definition* of thermodynamic entropy, say, is not really well-formed. In the context of limiting intertheoretic relations, the study of the thermodynamic limit renders such a question moot.

I will say more about this shortly. But first we must address some obvious questions and objections to the argument presented above. On the one hand, Sinai's demonstration of the emergence and ubiquity of the Gaussian demonstration assumes that the random variables are independent. To what extent is this assumption legitimate for real thermodynamic systems? On the other hand, it also assumes that we are working with a particular subsequence of the integers, $n_p = 2^p$. What is the role of this assumption? Finally, how does this entire argument, one which is purely mathematical, relate to real systems?

Let us take up the second question first. It is absolutely crucial. The particular subsequence $n_p = 2^p$ was chosen so that we would have the right normalization factor for the sum function – for the random variable $S_p$. One of Kadanoff's main insights was that as we sum over blocks of spins (which is, in effect, what we are doing by considering sum functions) and then try to replace these sums with individual block-spins (averages perhaps) we need to try to keep the (Hamiltonian) structure looking the same at these different scales. Equivalently, we need to make sure that the partition function for the original lattice is similar to that for the lattice of block-spins. One of the main difficulties of employing the renormalization techniques in investigating the behaviour of model systems is to determine exactly how one needs to rescale (or normalize) a collective variable like $\sum S_i$ so as to arrive at reasonable results. In fact, this is largely the art of the renormalization group approach. In the example above, this difficult problem is taken care of by employing that very special subsequence of the integers to form our 'blocks' of spins. This allows us to write down, without any difficulty, the 'renormalization group equation', (9.12), for our distribution functions. In effect, the use of the special subsequence of the integers gives us the proper normalization for the variables $S_p$ and determines the factor $\sqrt{2}$ that appears in Equation (9.12). In physical applications, however, we do not have the luxury of beginning with a solution to how the block-spins are to be normalized.

The first problem concerns the fact that the demonstration of the universality of the Gaussian distribution relies upon the fact that the random variables considered are independently distributed. Of course, this is far from true even for the simplest of thermodynamic systems: For example, the kinetic energies of molecules in an isolated gas are correlated with one another as a result of collisions and the fixed total energy of the gas. This poses a problem for Khinchin's program – one of which he was well-aware. He tries to deal with it by insisting that we must really think of the particles (in a real gas) as being 'only approximately isolated energetically components'. When being precise, however, we must allow for correlations between the components that would, strictly speaking, block the ability to use the central limit theorem. He says,

> inasmuch as forces of interaction between particles manifest themselves only at very small distances, such mixed terms in the expression of energy, representing mutual potential energy of particles, will be (in the great majority of points of the phase space) negligible as compared with the kinetic energy of particles or with the potential energy of external fields. In particular, they will contribute very little in evaluating various averages.... However, these mixed terms that are neglected, from the point of principle play a very important role, since it is precisely their presence that assures the possibility of an exchange of energy between the particles, on which is based the whole of statistical mechanics.
> (Khinchin, 1949: 43–43)

This is hardly a satisfying response to the problem of interaction or correlation: 'The entire success of statistical mechanics depends upon such interactions, but we *need* to ignore them completely because they will be small'.

In fact, though, it is possible to generalize the above argument for the universality of the Gaussian distribution to sequences of *dependent* random variables, provided that the dependence is not too strong. This is a truly remarkable feature of the central limit theorem. One can show that limiting Gaussian behaviour is to be expected even for a large class of *dependent* random variables. Without going into details, let me discuss the degree of robustness of such central limiting behaviour and when one might actually expect it to fail. To this end, consider once again a sequence of independent and identically distributed random variables $S_i$ ($i = 1, 2, \ldots$).[5] Suppose the $S_i$'s are centred at the origin and have finite variance. This means, respectively, that

$$E(S_i) = 0$$
$$E(S_i^2) = \sigma^2$$

Consider the sum function $S_N = \sum_{i=1}^{N} S_i$. The mean square deviation of $S_N$ is given by

$$E(S_N^2) = \sum_{i=1}^{N} E(S_i^2) = N\sigma^2 \qquad (9.13)$$

If we now suppose that the variables $S_i$ are not independent (9.13) needs to be modified to reflect the contribution of correlations to this average. So write

$$E(S_N^2) = \sum_{i,j=1}^{N} E(S_i S_j) = \sum_{i,j=1}^{N} R(i, j) \qquad (9.14)$$

---

[5] Here I follow the discussion in Cassandro and Jona-Lasinio (1978: 914–916).

where $R(i, j)$ is the correlation function of the sequence. In physical applications we typically consider only stationary sequences which means that there is a kind of translation invariance – no privileged $i$ in the sequence. Assuming stationarity we get[6]

$$E(S_N^2) = \sum_{i,j=1}^{N} R(i, j) = NR(0) + 2\sum_{l=1}^{N}(N - l)R(l) \qquad (9.15)$$

The first term reduces to $N\sigma^2$ when the variables are independent and so represents the contributions to the fluctuations of the variable $S_N$ if the sequence had been independent with variance $R(0)$. The second term is non-zero when correlations between the $S_i$ exist. Cassandro and Jona-Lasinio consider several possibilities for the behaviour of $R(l)$ (Cassandro and Jona-Lasinio, 1978; 915). The two most relevant to our discussion are the following:

$$\sum_{l=1}^{N} R(l) \xrightarrow{N \to \infty} \bar{R} < \infty \qquad (9.16)$$

and

$$\sum_{l=1}^{N} R(l) \xrightarrow{N \to \infty} \infty \qquad (9.17)$$

When (9.16) holds it is clear that for large $N$, the mean square deviation of $S_N$, $E(S_N^2)$, will still be proportional to $N$ for large $N$ and the $\sqrt{N}$ fluctuation law will still hold. Cassandro and Jona-Lasinio say that in this situation the variables $S_i$ are 'weakly dependent'. Weakly dependent systems of variables satisfy the condition of 'strong mixing' that, physically, is related to an exponential decay of correlations with distance. *In such cases it is possible to show that the distribution functions of these weakly dependent random variables will flow, under the transformation T to the Gaussian fixed point.* Thus, this answers the worry about using the central limit theorem in situations where independence fails. That is to say, central limiting behaviour is quite robust and the independence condition in the theorem as it is usually formulated can be considerably weakened. There is, therefore, no need to accept Khinchin's somewhat handwaving and inconsistent solution to the problem of interacting components.

When (9.17) holds

$$\frac{E(S_N^2)}{N} \to \infty$$

---

[6] Stationarity entails that $l$ will be the same for any pairs $(i, j)$ with equal separation ($|i - j|$).

for large $N$ and the fluctuations will be greater than $\sqrt{N}$. This is the situation of 'strongly dependent' variables. Now if we think of the variables $S_i$ as spins on a $d$-dimensional lattice $\mathbb{Z}^d$, we expect stationarity to hold which is to say that $R(\mathbf{i}, \mathbf{j}) = \mathbf{R}(\mathbf{i} - \mathbf{j})$ where $\mathbf{i}$ and $\mathbf{j}$ are $d$-dimensional lattice vectors. In such a case, (9.17) is expressed as follows:

$$\sum_{l \in \mathbb{Z}^d} R(l) = \infty \tag{9.18}$$

This says that the correlation length between spins on the lattice is *infinite*. In statistical mechanics this means that the system is at a point of second order phase transition – a critical point. This takes us directly to what very well may be another fundamental reason for Gibbs' caution in identifying thermodynamic quantities with (statistical) mechanical features of systems – the existence of critical phenomena.

Before turning to this let me briefly take stock of the status of reductive relations between thermodynamics and statistical mechanics up to this point. The argument of this section is designed to show that, *pace* Gibbs, we should not really worry about the fact that different 'precisely defined quantities' have some kind of equal claim to be 'regarded as temperature and entropy with respect to systems of a finite number of degrees of freedom' (Gibbs, 1981: 169). In fact, the canonical and micro-canonical distributions will yield the same results in the thermodynamic limit. They both lie within the basin of attraction of the Gaussian distribution – in the Gaussian universality class – under the probabilistic version of the renormalization group transformation sketched above. So Gibbs is being overly cautious from the point of view being advocated here.

By considering the thermodynamic limit in the context of intertheoretic relations, we do in effect demonstrate the existence of a limiting connection between statistical mechanics and thermodynamics akin to the physicists' schema (9.5) for reduction. That is to say that schema (9.5) holds where $T_f$ is statistical mechanics and $T_c$ is thermodynamics and $\epsilon = 1/N$.[7] The proof of the equivalence of the micro-canonical and the canonical ensembles in the thermodynamic limit, together with the probabilistic renormalization group argument given above, provides the required limiting connections between the two theories. Given this, we should expect something like a philosopher's derivational reduction to obtain. However, contrary to the Nagelian point of view which requires bridge laws identifying a thermodynamic quantity with a *unique* statistical mechanical quantity, we instead are able only to provide an association between the thermodynamical properties,

---

[7] More precisely, the thermodynamic limit is the limit in which $N \to \infty$, the volume $V \to \infty$ with the constraint that the density $N/V \to$ constant.

such as temperature and entropy, and a *universality class* of statistical mechanical structures.

## 9.8 Critical phenomena: the *inequivalence* of ensembles

While we seem to have been successful in finding a limiting connection between statistical mechanics and thermodynamics, there are *prima facie* reasons to believe that such a reduction cannot hold in general. This is because, the limiting schema (9.5) fails to hold for every thermodynamic phase. The existence of phase transitions and critical points guarantee that a smooth limiting relationship between the theories cannot hold everywhere.

Physically this failure of smooth limiting behaviour is related to the divergence of various quantities at the critical points. For instance, at a critical point $(P_c, V_c, T_c)$ the compressibility

$$\kappa \equiv \left[ -V \left( \frac{\partial P}{\partial V} \right)_T \right]^{-1} \tag{9.19}$$

becomes infinite.

From a modern perspective this failure can be related to the *inequivalence* of the micro-canonical and canonical ensembles at certain phases. Giovanni Gallavotti takes this to be a virtue of statistical mechanics rather than a defect.

> ... *Rather than being an obstacle* to the microscopic formulation of thermodynamics, [this inequivalence] *shows* the possibility that statistical mechanics can be a *natural frame* in which to study the *phase transition* phenomenon....
>
> We can therefore conclude, in the case ... of the canonical and microcanonical ensembles, that they provide equivalent descriptions of the system thermodynamics in the correspondence of the parameter values to which no phase transition is associated. In the other cases, the possible nonequivalence [inequivalence] *cannot be considered a defect of the theory*, but it can be ascribed to the fact that, when equivalence fails, the elements of the two statistical ensembles that should be equivalent are not because they describe two different phases that may coexist (or different mixtures of coexisting phases).
>
> *(Gallavotti, 1999: 74–75)*

Now Gibbs' work on thermodynamics, prior to writing *Elementary Principles in Statistical Mechanics*, focused largely on representing (geometrically and analytically) different phases of thermodynamic systems including places of phase transitions and critical points.[8] It is inconceivable to me that he was not aware of the problem that the existence of critical states raises for identifying precisely defined (statistical) mechanical properties with thermodynamic properties. That

---

[8] See, for example Gibbs (1928) and other papers in Volume One of his *Collected Works*.

is, I think one major factor behind Gibbs' caution – a major reason he allowed himself to speak *only* of thermodynamic analogies and not of identities – was the recognition that such *identities* will clearly fail at critical phases.

It is, therefore, extremely odd that Gibbs makes no mention of phase transitions and critical phenomena in *Elementary Principles in Statistical Mechanics*. This omission has been noted elsewhere by A. S. Wightman. Wightman states that

> there is one aspect of the thermodynamic limit that Gibbs does not emphasize. That is the appearance of phase transitions between distinct thermodynamic phases. Such sharp phase transitions do not occur in finite systems.... It is only in the thermodynamic limit that sharp phase transitions appear. A little more pedagogical zeal by Gibbs on this point would have saved some of the generations that followed considerable time.
> 
> *(Wightman, 1989: 34)*

It is certainly true that one can only find sharp (non-analytic) phase transitions in the thermodynamic limit. It is also true that that limit is singular on lines of phase transitions and at critical points.

As Gallavotti has noted, the equivalence of the ensembles in the thermodynamic limit can fail, and when it does, it means that the system is at criticality. We can no longer expect Gaussian behaviour to emerge as a limiting distribution in the iterative renormalization group Equation (9.12). Some have argued that this situation is an indication that we cannot expect general reductive limiting relations between thermodynamics and statistical mechanics to obtain.[9] However, if we take the inequivalence of the ensembles, and the existence of critical phenomena, to be two sides of the same coin, as Gallavotti suggests, then perhaps this pessimistic assessment is too quick. In this section, I would like to suggest that the probabilistic interpretation of the renormalization group provides a mathematical framework with which to investigate both sides of this coin using the same methodology. We shall see that the divergences and singularities at critical phases are not genuine obstacles to some kind of general limiting (reductive?) relation between the theories after all. Rather than give up on intertheoretic reduction in this case, we just need to recognize how involved it really is, and how different the reductive relation really is from the traditional philosophical conception.

We have seen that Gaussian limiting behaviour fails when the random variables $S_i$ are *strongly dependent*. This is the situation expressed by Equation (9.17), generally, and by Equation (9.18) for the case of spins on a $d$-dimensional lattice. Equivalently, given Equation (9.14), this means that the variance of the distribution functions for the sum functions or block-spins is infinite. Consider a generalization

---

[9] I, for one, have been a major proponent of this point of view.

(one that takes correlations into consideration) of the iterative scheme (9.11):[10]

$$F_{p+1} = g_N(x, a) \int_{-\infty}^{\infty} F_p(a^{-1}x - u) \, dF_p(u) \tag{9.20}$$

When the variables are strongly dependent $g_N(x, a) \to g_\infty(x, a)$ as $N \to \infty$ and this equation will, in that limit, have *non-Gaussian* solutions. That is, there will be limiting distributions $F^*$ satisfying the fixed point equation

$$TF^* = F^*$$

for the relevant renormalization group transformation analogous to (9.12):

$$TF = g_\infty(x, a) \int_{-\infty}^{\infty} F_\infty(a^{-1}x - u) \, dF_\infty(u) \tag{9.21}$$

Under the assumption that a fixed point solution to this equation exists, the value of the parameter $a$ will determine the critical exponent $\alpha \neq 1/2$ that characterizes the nature of this non-Gaussian limiting distribution as well as the form of the required normalization. If we assume that the correlation $R(l)$ exhibits power law behaviour $R(l_\alpha) \sim l^{-\alpha}$ and that $\sum_{l=1}^{N} R(l) \sim \sum_{l=1}^{N} l^{-\alpha} \to \infty$ as $N \to \infty$, then from Equation (9.15) we have

$$E(S_N^2) \to N^{2-\alpha}$$

as $N \to \infty$.

The point of all of this is the following: There *do* exist non-Gaussian probability distributions that are stable limit distributions under transformations of the form (9.21).[11] These distributions represent fixed points of the probabilistic renormalization group transformation for strongly dependent random variables. Strong dependence is exactly what we expect in real systems at criticality: It is reflected in the divergence of the correlation length. By generalizing the argument of the previous section, it is possible to show that these fixed points have large basins of attraction reflecting the fact that at criticality a wide variety of distributions will all exhibit the same, *non-Gaussian*, behaviour in the limit $N \to \infty$.

In effect, this solves the critical phenomenon problem. We can explain the universality (the virtual independence of behaviour from microscopic detail) of systems at criticality. The *very same strategy* that can be used to show that the micro-canonical and canonical distributions behave like Gaussians in the thermodynamic limit is used to explain the universal non-Gaussian behaviour of systems at criticality.

---

[10] The assumption that something like the convolution equation for distribution function of a sum of two random variables holds, is the essence of the renormalization group approach.

[11] Stability here means that a block random variable (a block-spin or sum function) with distribution function $F$ can be split into the sum of two block random variables of arbitrary relative size having a probability distribution of the same type; namely, $F$. Recall Equation (9.7).

## 9.9 Conclusion

First, a note about my use of the terms 'non-equivalence' and 'inequivalence' of ensembles. I use 'non-equivalence' in the context of Gibbs' recognition that the finite exact quantities calculated with respect to the micro-canonical and canonical ensembles are different. As noted, he took this to be a reason for scepticism about having found the right statistical mechanical quantity to represent the relevant thermodynamic quantity. I use 'inequivalence' in the context of infinite systems (that is, in the thermodynamic limit) where, for critical phases, the micro-canonical and canonical ensembles yield different results.

Second, I have argued that the best understanding of intertheoretic relations between statistical mechanics and thermodynamics is to be had by investigating the limiting relations, both regular and singular that can be expressed in terms of schema (9.5). We can say (with various qualifications[12]) that away from critical points and phase transitions, statistical mechanics does reduce to thermodynamics in the thermodynamic limit. And, we can say that there does exist a kind of identification of certain thermodynamic quantities such as temperature and entropy with *universality classes* of statistical mechanical quantities. In this context – that is, from the modern perspective of the success of renormalization group arguments – we see that Gibbs need not have been so cautious regarding the status of the connections between mechanical quantities and thermodynamic quantities. The existence of the thermodynamic limit, and the demonstration of the equivalence of ensembles, provides evidence that the question of which ensemble quantity is *really* to be identified with thermodynamic entropy, say, may not even be an important question to ask.

On the other hand, Gibbs was an expert on phase transitions and critical phenomena. I think that he must have understood that such phenomena present a problem for connecting the two theories. In his interest to avoid saying anything false, he did not raise the issue in his book. From the point of view I have been advocating, his understanding this problem is tantamount to recognizing that the limiting relationship between statistical mechanics and thermodynamics as $1/N \to 0$ is not everywhere regular. Despite this, the framework provided by the probabilistic interpretation of renormalization group argument indicates that it is possible to employ these techniques to connect both non-critical and critical thermodynamics to an underlying statistical mechanics. Both problems – the equivalence of ensembles and the inequivalence of ensembles – receive a unified, coherent treatment.

---

[12] These are, in fact, serious qualifications that may lead us ultimately to deny anything like a Nagelian reduction of the full theory of thermodynamics to statistical mechanics. See Sklar (1993: Chapter 9).

## References

Batterman, R. W. (1995) Theories between theories: asymptotic limiting intertheoretic relations. *Synthese*, **103**, 171–201.

Batterman, R. W. (2002a). Asymptotics and the role of minimal models. *The British Journal for the Philosophy of Science*, **53**, 21–38.

Batterman, R. W. (2002b). *The Devil in the Details: Asymptotic Reasoning in Explanation, Reduction, and Emergence. Oxford Studies in Philosophy of Science*. Oxford: Oxford University Press.

Bleher, P. M. and Sinai, Ya. G. (1973). Investigation of the critical point in models of the type of Dyson's hierarchical models. *Communications in Mathematical Physics*, **33**, 23–423.

Cassandro, M. and Jona-Lasinio, G. (1978). Critical point behaviour and probability theory. *Advances in Physics*, **27**, 913–941.

Gallavotti, G. (1999). *Statistical Mechanics: a Short Treatise. Texts and Monographs in Physics*. Berlin: Springer-Verlag.

Gibbs, J. W. (1928). A method of geometrical representation of the thermodynamic properties of substances by means of surfaces. In *The Collected Works of J. Willard Gibbs*, ch. II. New York: Longmans, Green and Co., pp. 33–54. (Originally published in (1893), *Transactions of the Connecticut Academy, II*, pp. 382–404)

Gibbs, J. W. (1981). *Elementary Principles in Statistical Mechanics: Developed with Especial Reference to the Rational Foundation of Thermodynamics*. Woodbridge, CN: Ox Bow Press. (First Published in 1902.)

Goldenfeld, N. (1992). *Lectures on Phase Transitions and the Renormalization Group. Frontiers in Physics*, no. 85 Reading, MA: Addison-Wesley.

Jona-Lasinio, G. (1975). The renormalization group: a probabilistic view. *Il Nuovo Cimento B*, **26**(1), 99–119.

Khinchin, A. I. (1949). *Mathematical Foundations of Statistical Mechanics*. New York: Dover Publications.

Nagel. E. (1961). *The Structure of Science: Problems in the Logic of Scientific Explanation*. New York: Harcourt, Brace, & World.

Sinai, Ya. G. (1978). Mathematical foundations of the renormalization group method in statistical physics. In *Mathematical Problems in Theoretical Physics*. S. Doplicher, G. Dell'Antonio and G. Jona-Lasinio, Berlin: Springer-Verlag, pp. 303–311.

Sinai, Ya. G. (1982). *Theory of Phase Transitions: Rigorous Results*. Oxford: Pergamon Press.

Sinai, Ya. G. (1992). *Probability Theory: an Introductory Course*, transl. by D. Haughton. Berlin: Springer-Verlag.

Sklar, L. (1993). *Physics and Chance: Philosophical Issues in the Foundations of Statstical Mechanics*. Cambridge: Cambridge University Press.

Wightman, A. S. (1989). On the prescience of J. Willard Gibbs. In *Proceedings of the Gibbs Symposium: Yale University*, May 15–17, 1989, G. D. Mostow and D. G. Caldi, ed. New York: American Mathematical Society, American Institute of Physics.

# 10

# Irreversibility in stochastic dynamics*

JOS UFFINK

## 10.1 Introduction

Over recent decades, some approaches to non-equilibrium statistical mechanics, that differ decidedly in their foundational and philosophical outlook, have nevertheless converged in developing a common unified mathematical framework. I will call this framework 'stochastic dynamics', since the main characteristic feature of the approach is that it characterizes the evolution of the state of a mechanical system as evolving under stochastic maps, rather than under a deterministic and time-reversal invariant Hamiltonian dynamics.[1]

The motivations for adopting this stochastic type of dynamics come from at least three different viewpoints.

(1) 'Coarse graining' (cf. van Kampen, 1962; Penrose, 1970): In this view one assumes that on the microscopic level the system can be characterized as a (Hamiltonian) dynamical system with deterministic time-reversal invariant dynamics. However, on the macroscopic level, one is only interested in the evolution of macroscopic states, i.e. in a partition (or coarse graining) of the microscopic phase space into discrete cells. The usual idea is that the form and size of these cells are chosen in accordance with the limits of our observational capabilities.

On the macroscopic level, the evolution now need no longer be portrayed as deterministic. When only the macro-state of a system at an instant is given, it is in general not fixed what its later macro-state will be, even if the underlying microscopic evolution is deterministic. Instead, one can provide *transition probabilities*, that specify how probable the transition from any given initial macro-state to later macro-states is. Although it is impossible, without further assumptions, to say anything general about the evolution of the macroscopically characterized states, it is

---

* This chapter is an excerpt of a longer discussion in (Uffink 2007).
[1] Also, the name has already been used for precisely this approach by Sudarshan and coworkers, cf. (Sudarshan *et al.* 1961, Mehra and Sudarshan 1972).

possible to describe the evolution of an ensemble or a probability distribution over these states, in terms of a *stochastic process*.

(2) 'Interventionism' or 'open systems' (cf. Blatt, 1959; Davies, 1976; Lindblad, 1976, Lindblad, 1983; Ridderbos, 2002): on this view, one assumes that the system to be described is not isolated but in (weak) interaction with the environment. It is assumed that the total system, consisting of the system of interest and the environment can be described as a (Hamiltonian) dynamical system with a time-reversal invariant and deterministic dynamics. If we represent the state of the system by $x \in \Gamma^{(s)}$ and that of the environment by $y \in \Gamma^{(e)}$, their joint evolution is given by a one-parameter group of evolution transformations, generated from the Hamiltonian equations of motion for the combined system: $U_t : (x, y) \mapsto U_t(x, y) \in \Gamma^{(s)} \times \Gamma^{(e)}$. The evolution of the state $x$ in the course of time is obtained by projecting, for each $t$, to the coordinates of $U_t(x, y)$ in $\Gamma^{(s)}$; call the result of this projection $x_t$. Clearly, this reduced time evolution of the system will generally fail to be deterministic, e.g. the trajectory described by $x_t$ in $\Gamma^{(s)}$ may intersect itself. Again, we may hope that this indeterministic evolution can nevertheless, for an ensemble of the system and its environment, be characterized as a stochastic process, at least if some further reasonable assumptions are made.

(3) A third viewpoint is to deny (Mackey, 1992; 2001), or to remain agnostic about (Streater, 1995), the existence of an underlying deterministic or time-reversal invariant dynamics, and simply regard the evolution of a system as described by a stochastic process as a new fundamental form of dynamics in its own right.

While authors in this approach thus differ in their motivation and in the interpretation they have of its subject field, there is a remarkable unity in the mathematical formalism adopted for this form of non-equilibrium statistical mechanics. The hope, obviously, is to arrange this description of the evolution of mechanical systems in terms of a stochastic dynamics in such a way that the evolution will typically display 'irreversible behaviour': i.e. an 'approach to equilibrium'; that a Boltzmann-like evolution equation holds; that there is a stochastic analogy of the *H*-theorem, etc. In short, one would like to recover the autonomy and irreversibility that thermal systems in non-equilibrium states typically display.

We will see that much of this can apparently be achieved with relatively little effort once a crucial technical assumption is in place: that the stochastic process is in fact a homogeneous Markov process, or, equivalently, obeys a so-called master equation. Much harder are the questions of whether the central assumptions of this approach might still be compatible with an underlying deterministic time-reversal invariant dynamics, and in which sense the results of the approach embody time-asymmetry. In fact we shall see that conflicting intuitions on this last issue arise, depending on whether one takes a probabilistic or a dynamics point of view towards this formalism.

From a foundational point of view, stochastic dynamics promises a new approach to the explanation of irreversible behaviour that differs in interesting ways from the more orthodox Hamiltonian or dynamical systems approach. In that approach, any account of irreversible phenomena can only proceed by referring to special initial conditions or dynamical hypotheses. Moreover, it is well-known that an ensemble of such systems will conserve (fine-grained) Gibbs entropy so that the account cannot rely on this form of entropy for a derivation of the increase of entropy. In stochastic dynamics, however, one may hope to find an account of irreversible behaviour that is not tied to special initial conditions, but one that is, so to say, built into the very stochastic-dynamical evolution. Further, since Liouville's theorem is not applicable, there is the prospect that one can obtain a genuine increase of Gibbs entropy from this type of dynamics.

As just mentioned, the central technical assumption in stochastic dynamics is that the processes described have the Markov property.[2] Indeed, general aspects of irreversible behaviour pour out almost effortlessly from the Markov property, or from the closely connected 'master equation'. Consequently, much of the attention in motivating stochastic dynamics has turned to the assumptions needed to obtain this Markov property, or slightly more strongly, to obtain a non-invertible Markov process (Mackey, 1992). The best-known specimen of such an assumption is van Kampen's (1962) 'repeated randomness assumption'. And similarly, critics of this type of approach (Sklar, 1993; Redhead, 1995; Callender, 1999) have also focused their objections on the question just how reasonable and general such assumptions are (cf. Section 10.5).

I believe both sides of the debate have badly missed the target. Many authors have uncritically assumed that the assumption of a (non-invertible) Markov process does indeed lead to non-time-reversal-invariant results. As a matter of fact, however, the Markov property (for invertible or non-invertible Markov processes) is time-reversal invariant. So, any argument to obtain that property need not presuppose time-asymmetry. In fact, I will argue that this discussion of irreversible behaviour as derived from the Markov property suffers from an illusion. It is due to the habit of studying the prediction of future states from a given initial state, rather than studying retrodictions towards an earlier state. As we shall see, for a proper description of irreversibility in stochastic dynamics one needs to focus on another issue, namely the difference between backward and forwards transition probabilities.

In the following sections, I will first (Section 10.2) recall the standard definition of a homogeneous Markov process from the theory of stochastic processes. Section 10.3 casts these concepts in the language of dynamics, introduces the

---

[2] Some authors argue that the approach can and should be extended to include non-Markovian stochastic processes as well. Nevertheless I will focus here on Markov processes.

master equation, and discusses its analogy to the Boltzmann equation. In Section 10.4, we review some of the results that *prima facie* display irreversible behaviour for homogeneous Markov processes. In Section 10.5 we turn to the physical motivations that have been given for the Markov property, and their problems, while Section 10.6 focuses on the question how seemingly irreversible results could have been obtained from time-symmetric assumptions. Finally, Section 10.7 argues that a more promising discussion of these issues should start from a different definition of reversibility of stochastic processes.

## 10.2 The definition of Markov processes

To start off, consider a paradigmatic example. One of the oldest discussions of a stochastic process in the physics literature is the so-called 'dog flea model' of P. and T. Ehrenfest (1907).

Consider $N$ fleas, labelled from 1 to $N$, situated on either of two dogs. The number of fleas on dog 1 and 2 are denoted as $n_1$ and $n_2 = N - n_1$. Further, we suppose there is an urn with $N$ lots carrying the numbers $1, \ldots, N$ respectively. The urn is shaken, a lot is drawn (and replaced), and the flea with the corresponding label is ordered to jump to the other dog. This procedure is repeated every second.

It is not hard to see that this model embodies an 'approach to equilibrium' in some sense: Suppose that initially all or almost all fleas are on dog 1. Then it is very probable that the first few drawings will move fleas from dog 1 to 2. But as soon as the number of fleas on dog 2 increases, the probability that some fleas will jump back to dog 1 increases too. The typical behaviour of, say, $|n_1 - n_2|$ as a function of time will be similar to Boltzmann's $H$-curve, with a tendency of $|n_1 - n_2|$ to decrease if it was initially large, and to remain close to the 'equilibrium' value $n_1 \approx n_2$ for most of the time. But note that in contrast to Boltzmann's $H$-curve in gas theory, the 'evolution' is here entirely stochastic, i.e. generated by a lottery, and that no underlying deterministic equations of motion are provided.

Generally speaking, a stochastic process is defined as a probability measure $P$ on a measure space $X$, whose elements will here be denoted as $\xi$, on which there are infinitely many random variables $Y_t$, with $t \in \mathbb{R}$ (or sometimes $t \in \mathbb{Z}$). Physically speaking, we interpret $t$ as time, and $Y$ as the macroscopic variable(s) characterizing the macro-state – say the number of fleas on a dog, or the number of molecules with their molecular state in some cell of $\mu$-space, etc. Further, $\xi$ represents the total history of the system which determines the values of $Y_t(\xi)$. The collection $Y_t$ may thus be considered as a single random variable $Y$ evolving in the course of time.

At first sight, the name 'process' for a probability measure may seem somewhat unnatural. From a physical point of view it is the *realization*, in which the random variables $Y_t$ attain the values $Y_t(\xi) = y_t$ that should be called a process. In the mathematical literature, however, it has become usual to denote the measure that determines the probability of all such realizations as a 'stochastic process'.

For convenience we assume here that the variables $Y_t$ may attain only finitely many discrete values, say $y_t \in \mathcal{Y} = \{1, \ldots, m\}$. However, the theory can largely be set up in complete analogy for continuous variables.

The probability measure $P$ provides, for $n = 1, 2, \ldots$, and instants $t_1, \ldots, t_n$ definite probabilities for the event that $Y_t$ at these instants attains certain values $y_1, \ldots, y_n$:

$$P_{(1)}(y_1, t_1)$$
$$P_{(2)}(y_2, t_2; y_1, t_1)$$
$$\vdots$$
$$P_{(n)}(y_n, t_n; \ldots; y_1, t_1) \qquad (10.1)$$
$$\vdots$$

Here, $P_{(n)}(y_n, t_n; \ldots; y_1, t_1)$ stands for the joint probability that at times $t_1, \ldots, t_n$ the quantities $Y_t$ attain the values $y_1, \ldots, y_n$, with $y_i \in \mathcal{Y}$. It is an abbreviation for

$$P_{(n)}(y_n, t_n; \ldots; y_1, t_1) := P(\{\xi \in X : Y_{t_n}(\xi) = y_n \ \& \ \cdots \ \& \ Y_{t_1}(\xi) = y_1\}) \qquad (10.2)$$

Obviously the probabilities (10.1) are normalized and non-negative, and each $P_{(n)}$ is a marginal of all higher-order probability distributions:

$$P_{(n)}(y_n, t_n; \ldots; y_1, t_1) = \sum_{y_{n+m}} \cdots \sum_{y_{n+1}} P_{(n+m)}(y_{n+m}, t_{n+m}; \ldots; y_1, t_1) \qquad (10.3)$$

In fact, the probability measure $P$ is uniquely determined by the hierarchy (10.1).[3]

Similarly, we may define conditional probabilities in the familiar manner, e.g.:

$$P_{(1|n-1)}(y_n, t_n | y_{n-1}, t_{n-1}; \ldots; y_1, t_1) := \frac{P_{(n)}(y_n, t_n; \ldots; y_1, t_1)}{P_{(n-1)}(y_{n-1}, t_{n-1}; \ldots; y_1, t_1)} \qquad (10.4)$$

provides the probability that $Y_{t_n}$ attains the value $y_n$, under the condition that $Y_{t_{n-1}}, \ldots, Y_{t_1}$ have the values $y_{n-1}, \ldots, y_1$. In principle, the times appearing in the joint and conditional probability distributions (10.1) and (10.4) may be chosen in an arbitrary order. However, we adopt from now on the convention that they are ordered as $t_1 < \cdots < t_n$.

A special and important type of stochastic process is obtained by adding the assumption that such conditional probabilities depend only the condition at the last

---

[3] At least, when we assume that the $\sigma$-algebra of measurable sets in $X$ is the cylinder algebra generated by sets of the form in the right-hand side of (10.2).

instant. That is to say: for all $n$ and all choices of $y_1, \ldots, y_n$ and $t_1 < \cdots < t_n$:

$$P_{(1|n)}(y_n, t_n | y_{n-1}, t_{n-1}; \ldots; y_n, t_n) = P_{(1|1)}(y_n, t_n | y_{n-1}, t_{n-1}) \qquad (10.5)$$

This is the *Markov property* and such stochastic processes are called *Markov processes*.

The interpretation often given to this assumption, is that Markov processes have 'no memory'. To explain this slogan more precisely, consider the following situation. Suppose we are given a piece of the history of the quantity $Y$: at the instants $t_1, \ldots, t_{n-1}$ its values have been $y_1, \ldots, y_{n-1}$. On this information, we want to make a prediction of the value $y_n$ of the variable $Y$ at a later instant $t_n$. The Markov-property (10.5) says that this prediction would not have been better or worse if, instead of knowing this entire piece of prehistory, only the value $y_{n-1}$ of $Y$ at the last instant $t_{n-1}$ had been given. Additional information about the past values is thus irrelevant for a prediction of the future value.

For a Markov process, the hierarchy of joint probability distributions (10.1) is subjected to stringent demands. In fact they are all completely determined by: (a) the specification of $P_{(1)}(y, 0)$ at one arbitrary chosen initial instant $t = 0$, and (b) the conditional probabilities $P_{(1|1)}(y_2, t_2 | y_1, t_1)$ for all $t_2 > t_1$. Indeed,

$$P_{(1)}(y, t) = \sum_{y_0} P_{(1|1)}(y, t | y_0, 0) P_{(1)}(y_0, 0) \qquad (10.6)$$

and for the joint probability distributions $P_{(n)}$ we find:

$$P_{(n)}(y_n, t_n; \ldots; y_1, t_1) = P_{(1|1)}(y_n, t_n | y_{n-1}, t_{n-1}) P_{(1|1)}(y_{n-1}, t_{n-1} | y_{n-2}, t_{n-2})$$
$$\times \cdots \times P_{(1|1)}(y_2, t_2 | y_1, t_1) P_{(1)}(y_1, t_1) \qquad (10.7)$$

It follows from the Markov property that the conditional probabilities $P_{(1|1)}$ have the following property, known as the *Chapman–Kolmogorov* equation:

$$P_{(1|1)}(y_3, t_3 | y_1, t_1) = \sum_{y_2} P_{(1|1)}(y_3, t_3 | y_2, t_2) P_{(1|1)}(y_2, t_2 | y_1, t_1) \text{ for } t_1 < t_2 < t_3$$
$$(10.8)$$

So, for a Markov process, the hierarchy (10.1) is completely characterized by specifying $P_{(1)}$ at an initial instant and a system of conditional probabilities $P_{(1|1)}$ satisfying the Chapman–Kolmogorov equation. The study of Markov processes therefore focuses on these two ingredients.[4]

---

[4] Note, however, that although every Markov process is fully characterized by (i) an initial distribution $P_{(1)}(y, 0)$ and (ii) a set of transition probabilities $P_{(1|1)}$ obeying the Chapman–Kolmogorov equation and the equations (10.7), it is *not* the case that every stochastic process obeying (i) and (ii) is a Markov process. (See van Kampen 1981: 83 for a counterexample.) Still, it is true that one can define a unique Markov process from these two ingredients by stipulating (10.7).

A Markov process is called *homogeneous* if the conditional probabilities $P_{(1|1)}(y_2, t_2|y_1, t_1)$ do not depend on the two times $t_1, t_2$ separately but only on their difference $t = t_2 - t_1$; i.e. if they are invariant under time translations. In this case we may write

$$P_{(1|1)}(y_2, t_2|y_1, t_1) = T_t(y_2, y_1) \qquad (10.9)$$

such conditional probabilities are also called *transition* probabilities.

Is the definition of a Markov process time-symmetric? The choice in (10.5) of conditionalizing the probability distribution for $Y_{t_n}$ on *earlier* values of $Y_t$ is of course special. In principle, there is nothing in the formulas (10.1) or (10.4) that forces such an ordering. One might, just as well, ask for the probability of a value of $Y_t$ in the past, under the condition that part of the *later* behaviour is given (or, indeed, conditionalize on the behaviour at both earlier and later instants.)

At first sight, the Markov property makes no demands about these latter cases. Therefore, one might easily get the impression that the definition is time-asymmetric. However, this is not the case. One can show that (10.5) is equivalent to:

$$P_{(1|n-1)}(y_1, t_1|y_2, t_2; \ldots; y_n, t_n) = P_{(1|1)}(y_1 t_1|y_2, t_2) \qquad (10.10)$$

where the convention $t_1 < t_2 < \cdots < t_n$ is still in force. Thus, a Markov process does not only have 'no memory' but also 'no foresight'.

Some authors (e.g. Kelly, 1979) adopt an (equivalent) definition of a Markov process that is explicitly time-symmetric: Suppose that the value $y_i$ at an instant $t_i$ somewhere in the middle of the sequence $t_1 < \cdots < t_n$ is given. The condition for a stochastic process to be Markov is then

$$P_{(n|1)}(y_n, t_n; \ldots; y_1, t_1|y_i, t_i) =$$
$$P_{(n-i|1)}(y_n, t_n; \ldots; y_{i+1}, t_{i+1}|y_i, t_i) \, P_{(i-1|1)}(y_{i-1}, t_{i-1}; y_1, t_1|y_i, t_i) \qquad (10.11)$$

for all $n = 1, 2, \ldots$ and all $1 \leq i \leq n$. In another slogan: The future and past are independent if one conditionalizes on the present.

## 10.3 Stochastic dynamics

A homogeneous Markov process is for $t > 0$ completely determined by the specification of an initial probability distribution $P_{(1)}(y, 0)$ and the transition probabilities $T_t(y_2|y_1)$ defined by (10.9). The difference in notation (between $P$ and $T$) also serves to ease a certain conceptual step. Namely, the idea is to regard $T_t$ as a stochastic evolution operator. Thus, we can regard $T_t(y_2|y_1)$ as the elements of a matrix, representing a (linear) operator $T$ that determines how an initial distribution

$P_{(1)}(y, 0)$ will evolve into a distribution at later instants $t > 0$. (In the sequel I will adapt the notation and write $P_{(1)}(y, t)$ as $P_t(y)$.)

$$P_t(y) = (T_t P)(y) := \sum_{y'} T_t(y|y') P_0(y') \tag{10.12}$$

The Chapman–Kolmogorov equation (10.8) may then be written compactly as

$$T_{t+t'} = T_t \circ T_{t'} \quad \text{for } t, t' \geq 0 \tag{10.13}$$

where $\circ$ stands for matrix multiplication, and we now also extend the notation to include the unit operator:

$$1(y, y') = T_0(y, y') := \delta_{y, y'} \tag{10.14}$$

where $\delta$ denotes the Kronecker delta.

The formulation (10.3) can (almost) be interpreted as the group composition property of the evolution operators $T$. It may be instructive to note how much this is due to the Markov property. Indeed, for arbitrary conditional probabilities, say, if $A_i$, $B_j$ and $C_k$ denote three families of complete and mutually exclusive events the rule of total probability gives:

$$P(A_i|C_k) = \sum_j P(A_i|B_j, C_k) P(B_j|C_k) \tag{10.15}$$

In general, this rule can *not* be regarded as ordinary matrix multiplication or a group composition! But the Markov property makes $P(A_i|B_j, C_k)$ in (10.15) reduce to $P(A_i|B_j)$, and then the summation in (10.15) coincides with the familiar rule for matrix multiplication.

I wrote above: 'almost', because there is still a difference in comparison with the normal group property: in the Chapman–Kolmogorov equation (10.13) all times must be positive. Thus, in general, for $t < 0$, $T_t$ may not even be defined and so it does *not* hold that

$$T_{-t} \circ T_t = 1 \tag{10.16}$$

A family of operators $\{T_t, t \geq 0\}$ which is closed under an operation $\circ$ that obeys (10.13), and for which $T_0 = 1$ is called a *semigroup*. It differs from a group in the sense that its elements $T_t$ need not be *invertible*, i.e. need not have an inverse. The lack of an inverse of $T_t$ may be due to various reasons: either $T_t$ does not possess an inverse, i.e. it is not a one-to-one mapping, or $T_t$ does possess an inverse matrix $T_t^{\text{inv}}$, which however is itself non-stochastic (e.g. it may have negative matrix-elements). We will come back to the role of the inverse matrices in Sections 10.4 and 10.7.

The theory of Markov processes has a strong and natural connection with linear algebra. Sometimes, the theory is presented entirely from this perspective, and one starts with the introduction of a semigroup of *stochastic matrices*, that is to say, $m$ by $m$ matrices $T$ with $T_{ij} \geq 0$ and $\sum_i T_{ij} = 1$. Or, more abstractly, one posits a class of states $P$, elements of a Banach space with a norm $\|P\|_1 = 1$, and a semigroup of stochastic maps $T_t$, ($t \geq 0$), subject to the conditions that $T_t$ is linear, positive, and preserves norm: $\|T_t P\|_1 = \|P\|_1$ (see Streater, 1995).

The evolution of a probability distribution $P$ (now regarded as a vector or a state) is then particularly simple when $t$ is discrete ($t \in \mathbb{N}$):

$$P_t = T^t P_0, \text{ where } T^t = \underbrace{T \circ \cdots \circ T}_{t \text{ times}} \tag{10.17}$$

Homogeneous Markov processes in discrete time are also known as *Markov chains*.

Clearly, if we consider the family $\{T_t\}$ as a semigroup of stochastic evolution operators, or a stochastic form of dynamics, it becomes attractive to look upon $P_0(y)$ as a contingent initial state, chosen independently of the evolution operators $T_t$. Still, from the perspective of the probabilistic formalism with which we started, this might be an unexpected thought: both $P_{(1)}$ and $P_{(1|1)}$ are aspects of a single, given, probability measure $P$. The idea of regarding them as independent ingredients that may be specified separately does not then seem very natural. But, of course, there is no formal objection against the idea, since every combination of a system of transition probabilities $T_t$ obeying the Chapman–Kolmogorov equation, and an arbitrary initial probability distribution $P_0(y) = P_{(1)}(y, 0)$ defines a unique homogeneous Markov process (cf. footnote 4). In fact, one sometimes even goes one step further and identifies a homogeneous Markov process completely with the specification of the transition probabilities, without regard of the initial state $P_0(y)$; just like the dynamics of a deterministic system is usually presented without assuming any special initial state.

For Markov chains, the goal of specifying the evolution of $P_t(y)$ is now already completely solved in Equation (10.17). In the case of continuous time, it is more usual to specify evolution by means of a differential equation. Such an equation may be obtained in a straightforward manner by considering a Taylor expansion of the transition probability for small times (van Kampen, 1981: 101–103) – under an appropriate continuity assumption.

The result (with a slightly changed notation) is:

$$\frac{\partial P_t(y)}{\partial t} = \sum_{y'} \left( W(y|y') P_t(y') - W(y'|y) P_t(y) \right) \tag{10.18}$$

Here, the expression $W(y|y')$ is the transition probability from $y'$ to $y$ *per unit of time*. This differential equation, first obtained by Pauli in 1928, is called the *master equation*.

The interpretation of the equation is suggestive: the change of the probability $P_t(y)$ is determined by making up a balance between gains and losses: the probability of value $y$ increases in a time $dt$ because of the transitions from $y'$ to $y$, for all possible values of $y'$. This increase per unit of time is $\sum_{y'} W(y|y')P_t(y')$. But in same period $dt$ there is also a decrease of $P_t(y)$ as a consequence of transitions from the value $y$ to all other possible values $y'$. This provides the second term.

In this 'balancing' aspect, the master equation resembles the Boltzmann equation, despite the totally different derivation, and the fact that $P_t(y)$ has quite another meaning than Boltzmann's $f_t(v)$. (The former is a probability distribution, the latter a distribution of particles.) Both are first-order differential equations in $t$. A crucial mathematical distinction from the Boltzmann equation is that the master equation is linear in $P$, and therefore much easier to solve.

Indeed, any solution of the master equation can formally be written as:

$$P_t = e^{tL} P_0 \qquad (10.19)$$

where $L$ represents the operator

$$L(y|y') := W(y|y') - \sum_{y''} W(y''|y')\delta_{y,y'} \qquad (10.20)$$

The general solution (10.19) is similar to the discrete time case (10.17), thus showing the equivalence of the master equation to the assumption of a homogeneous Markov process in continuous time.

## 10.4 Approach to equilibrium and increase of entropy?

What can we say in general about the evolution of $P_t(y)$ for a homogeneous Markov process? An immediate result is this: the *relative entropy* is monotonically non-decreasing. That is to say, if we define

$$H(P, Q) := -\sum_{y \in \mathcal{Y}} P(y) \ln \frac{P(y)}{Q(y)} \qquad (10.21)$$

as the relative entropy of a probability distribution $P$ relative to $Q$, then one can show (see e.g. Moran, 1961; Mackey, 1992: 30):

$$H(P_t, Q_t) \geq H(P, Q) \qquad (10.22)$$

where $P_t = T_t P$, $Q_t = T_t Q$, and $T_t$ are elements of the semigroup (10.17) or (10.19). One can also show that in the case of a non-zero relative entropy increase for at least some pair of probability distributions $P$ and $Q$, the stochastic matrix $T_t$ must be non-invertible.

The relative entropy $H(P, Q)$ can, in some sense, be thought of as a measure of how much $P$ and $Q$ 'resemble' each other.[5] Indeed, it takes its maximum value (i.e. 0) if and only if $P = Q$; it may become $-\infty$ if $P$ and $Q$ have disjoint support (i.e. when $P(y)Q(y) = 0$ for all $y \in \mathcal{Y}$). Thus, the result (10.22) says that if the stochastic process is non-invertible, pairs of distributions $P_t$ and $Q_t$ will generally become more and more alike as time goes by.

Hence it seems we have obtained a general weak aspect of 'irreversible behaviour' in this framework. Of course, the above result does not yet imply that the 'absolute' entropy $H(P) := -\sum_y P(y) \ln P(y)$ of a probability distribution is non-decreasing. But now assume that the process has a *stationary distribution*. In other words, there is a probability distribution $P^*(y)$ such that

$$T_t P^* = P^* \tag{10.23}$$

The intention is, obviously, to regard such a distribution as a candidate for the description of an equilibrium state. If there is such a stationary distribution $P^*$, we may apply the previous result and write:

$$H(P, P^*) \leq H(T_t P, T_t P^*) = H(P_t, P^*) \tag{10.24}$$

In other words, as time goes by, the distribution $T_t P$ will then more and more resemble the stationary distribution than does $P$. If the stationary distribution is also uniform, i.e.:

$$P^*(y) = \frac{1}{m} \tag{10.25}$$

then not only the relative but also the absolute entropy $H(P) := -\sum_y P(y) \ln P(y)$ increases, because

$$H(P, P^*) = H(P) - \ln m \tag{10.26}$$

In order to get a satisfactory description of an 'approach to equilibrium' the following questions remain:

(i) is there such a stationary distribution?
(ii) if so, is it unique?
(iii) does the monotonic behaviour of $H(P_t)$ imply that $\lim_{t \to \infty} P_t = P^*$?

Harder questions, which we postpone to the next Subsection 10.5, are:

(iv) how to motivate the assumptions needed in this approach or how to judge their (in)compatibility with an underlying time deterministic dynamics; and
(v) how this behaviour is compatible with the time-symmetry of Markov processes.

---

[5] Of course, this is an asymmetric sense of 'resemblance' because $H(P, Q) \neq H(Q, P)$.

## Ad (i)

A stationary distribution as defined by (10.23), can be seen as an eigenvector of $T_t$ with eigenvalue 1, or, in the light of (10.19), an eigenvector of $L$ for the eigenvalue 0. Note that $T$ or $L$ are not necessarily Hermitian (or, rather, since we are dealing with real matrices, symmetric), so that the existence of eigenvectors is not guaranteed by the spectral theorem. Further, even if an eigenvector with the corresponding eigenvalue exists, it is not automatically suitable as a probability distribution because its components might not be positive.

Still, it turns out that, due to a theorem of Perron (1907) and Frobenius (1909), every stochastic matrix indeed has an eigenvector, with exclusively non-negative components, and eigenvalue 1 (see e.g. Gantmacher, 1959). But if the set $\mathcal{Y}$ is infinite or continuous this is not always true.

However, for continuous variables with a range that has finite measure, the existence of a stationary distribution is guaranteed under the condition that the probability density $\rho_y$ is at all times bounded, i.e. $\exists M \in \mathbb{R}$ such that $\forall t\ \rho_t \leq M$; (see Mackey, 1992: 36).

## Ad (ii)

The question whether stationary distributions will be unique is somewhat harder to tackle. This problem exhibits an analogy to that of metric transitivity in the ergodic problem.

In general, it is very well possible that the range $\mathcal{Y}$ of $Y$ can be partitioned in two disjoint regions, say $A$ and $B$, with $\mathcal{Y} = A \cup B$, such that there are no transitions from $A$ to $B$ or vice versa (or that such transitions occur with probability zero). That is to say, the stochastic evolution $T_t$ might have the property

$$T_t(Y \in A | Y \in B) = T_t(Y \in B | Y \in A) = 0 \qquad (10.27)$$

In other words, its matrix may (perhaps after a conventional relabelling of the outcomes) be written in the form:

$$\begin{pmatrix} T_A & 0 \\ 0 & T_B \end{pmatrix} \qquad (10.28)$$

The matrix is then called (completely) *reducible*. In this case, stationary distributions will generally not be unique: If $P_A^*$ is a stationary distribution with support in the region $A$, and $P_B^*$ is a stationary distribution with support in $B$, then every convex combination

$$\alpha P_A^*(y) + (1-\alpha) P_B^*(y) \text{ with } 0 \leq \alpha \leq 1 \qquad (10.29)$$

will be stationary too. In order to obtain a unique stationary distribution we will thus have to assume an analogue of metric transitivity. That is to say: we should demand that every partition of $\mathcal{Y}$ into disjoint sets $A$ and $B$ for which (10.27) holds is 'trivial' in the sense that $P(A) = 0$ or $P(B) = 0$.

So, one may ask, is the stationary distribution $P^*$ unique if and only if the transition probabilities $T_\tau$ are not reducible? In the ergodic problem, the answer is positive (at least if $P^*$ is assumed to be absolutely continuous with respect to the micro-canonical measure). But not in the present case!

This has to do with the phenomenon of so-called 'transient states', which has no analogy in Hamiltonian dynamics. Let us look at an example to introduce this concept. Consider a stochastic matrix of the form:

$$\begin{pmatrix} T_A & T' \\ 0 & T_B \end{pmatrix} \tag{10.30}$$

where $T'$ is a matrix with non-negative entries only. Then:

$$\begin{pmatrix} T_A & T' \\ 0 & T_B \end{pmatrix} \begin{pmatrix} P_A \\ 0 \end{pmatrix} = \begin{pmatrix} T_A P_A \\ 0 \end{pmatrix}, \begin{pmatrix} T_A & T' \\ 0 & T_B \end{pmatrix} \begin{pmatrix} 0 \\ P_B \end{pmatrix} = \begin{pmatrix} T' P_B \\ T_B P_B \end{pmatrix} \tag{10.31}$$

so that here transitions of the type $a \longrightarrow b$ have probability zero, but transitions of the type $b \longrightarrow a$ occur with positive probability. (Here, $a, b$ stand for arbitrary elements of the subsets $A$ and $B$.) It is clear that in such a case the region $B$ will eventually be 'sucked empty'. That is to say: the total probability of being in region $B$ (i.e. $\|T^t P_B\|$) will go exponentially to zero. The distributions with support in $B$ are called 'transient' and the set $A$ is called 'absorbing' or a 'trap'.

It is clear that these transient states will not play any role in the determination of the stationary distribution, and that for this purpose they might be simply ignored. Thus, in this example, the only stationary distributions are those with a support in $A$. And there will be more than one of them if $T_A$ is reducible.

A matrix $T$ that may be brought (by permutation of the rows and columns) in the form (10.30), with $T_A$ reducible is called *incompletely reducible* (van Kampen, 1981: 108). Further, a stochastic matrix is called *irreducible* if it is neither completely or incompletely reducible. An alternative (equivalent) criterion is that all states 'communicate' with each other, i.e. that for every pair of $i, j \in \mathcal{Y}$ there is some time $t$ such that $P_t(j|i) > 0$.

The Perron–Frobenius theorem guarantees that as long as $T$ is irreducible, there is a unique stationary distribution. Furthermore, one can then prove an analogue of the ergodic theorem (Petersen, 1983: 52):

ERGODIC THEOREM FOR MARKOV PROCESSES: If the transition probability $T_t$ is irreducible, the time average of $P_t$ converges to the unique stationary solution:

$$\lim_{\tau \to \infty} \frac{1}{\tau} \int_0^\tau T_t P(y) \, dt = P^*(y) \qquad (10.32)$$

### Ad (iii)

If there is a unique stationary distribution $P^*$, will $T_t P$ converge to $P^*$, for every choice of $P$? Again, the answer is not necessarily affirmative. (Even if (10.32) is valid!) For example, there are homogeneous and irreducible Markov chains for which $P_t$ can be divided into two pieces: $P_t = Q_t + R_t$ with the following properties (Mackey, 1992: 71):

(1) $Q_t$ is a term with $\|Q_t\| \longrightarrow 0$. This is a transient term.
(2) The remainder $R_t$ is periodic, i.e. after some time $\tau$ the evolution repeats itself: $R_{t+\tau} = R_\tau$.

These processes are called *asymptotically periodic*. They may very well occur in conjunction with a unique stationary distribution $P^*$, and show strict monotonic increase of entropy, but still not converge to $P^*$. In this case, the monotonic increase of relative entropy $H(P_t, P^*)$ is entirely due to the transient term. For the periodic piece $R_t$, the transition probabilities are permutation matrices, which, after $\tau$ repetitions, return to the unit operator. Further technical assumptions can be introduced to block examples of this kind, and thus enforce a strict convergence towards the unique stationary distribution, e.g. by imposing a condition of 'exactness' (Mackey, 1992). However, it would take us too far afield to discuss this in detail.

In conclusion, it seems that a weak aspect of 'irreversible behaviour', i.e. the monotonic non-decrease of relative entropy is a general feature for all homogeneous Markov processes, (and indeed for all stochastic processes), and non-trivially so when the transition probabilities are non-invertible. Stronger versions of that behaviour, in the sense of affirmative answers to the questions (i), (ii) and (iii), can be obtained too, but at the price of additional technical assumptions.

## 10.5 Motivations for the Markov property and objections against them

### Ad (iv)

We now turn to the following problem: what is the motivation behind the assumption of the Markov property? The answer, of course, is going to depend on the interpretation of the formalism that one has in mind, and may be different in the 'coarse graining' and the 'open systems' or interventionist approaches (cf.

Section 10.1). I shall discuss the coarse-graining approach in the next section, and then consider the similar problem for the interventionist point of view.

### 10.5.1 Coarse graining and the repeated randomness assumption

In the present point of view, one assumes that the system considered is really an isolated Hamiltonian system, but the Markov property is supposedly obtained from a partitioning of its phase space. But exactly how is that achieved?

One of the clearest and most outspoken presentations of this view is (van Kampen, 1962). Assume the existence of some privileged partition of the Hamiltonian phase space $\Gamma$ – or of the energy hypersurface $\Gamma_E$ – into disjoint cells: $\Gamma = \omega_1 \cup \cdots \cup \omega_m$. Consider an arbitrary ensemble with probability density $\rho$ on this phase space. Its evolution can be represented by an operator

$$U_t^* \rho(x) := \rho(U_{-t}x) \tag{10.33}$$

where we use $U_t$ to denote the Hamiltonian evolution operators. Let transition probabilities between the cells of this partition be defined as

$$T_t(j|i) := P(x_t \in \omega_j | x \in \omega_i) = P(U_t x \in \omega_j | x \in \omega_i) = \frac{\int_{(U_{-t}\omega_j) \cap \omega_i} \rho(x)\,dx}{\int_{\omega_i} \rho(x)\,dx} \tag{10.34}$$

Obviously such transition probabilities will be homogeneous, due to the time-translation invariance of the Hamiltonian evolution $U_t$. Further, let $\hat{p}_0(i) := P(x \in \omega_i) = \int_{\omega_i} \rho(x)\,dx$, $i \in \mathcal{Y} = \{1, \ldots, m\}$, be an arbitrary initial coarse-grained probability distribution at time $t = 0$.

Using the coarse-graining map defined by:

$$\mathcal{CG} : \rho(x) \mapsto \mathcal{CG}\rho(x) = \sum_i \hat{\rho}(i) 1_{\omega_i}(x) \tag{10.35}$$

where

$$\hat{\rho}(i) := \frac{\int_{\omega_i} \rho(x)\,dx}{\int_{\omega_i} dx} \tag{10.36}$$

one may also express the coarse-grained distribution at time $t$ as

$$\mathcal{CG}U_t^* \rho(x) = \sum_{ji} T_t(j|i) \hat{p}_0(i) \frac{1}{\mu(\omega_j)} 1_{\omega_j}(x) \tag{10.37}$$

where $\mu$ is the canonical measure on $\Gamma$, or the micro-canonical measure on $\Gamma_E$. This expression indicates that, as long as we are only interested in the coarse-grained history, it suffices to know the transition probabilities (10.34) and the initial coarse-grained distributions.

But in order to taste the fruits advertised in the previous sections, one needs to show that the transition probabilities define a Markov process, i.e. that they obey the Chapman–Kolmogorov equation (10.8),

$$T_{t'+t}(k|i) = T_{t'}(k|j) T_t(j|i) \text{ for all } t, t' > 0 \tag{10.38}$$

Applying (10.37) for times $t$, $t'$ and $t+t'$, it follows easily that the Chapman–Kolmogorov equation is equivalent to

$$\mathcal{CG} U^*_{t'+t} = \mathcal{CG} U^*_{t'} \mathcal{CG} U^*_t \text{ for all } t, t' > 0 \tag{10.39}$$

In other words, the coarse-grained probability distribution at time $t+t'$ can be obtained by first applying the Hamiltonian dynamical evolution during a time $t$, then performing a coarse-graining operation, next applying the dynamical evolution during time $t'$, and then coarse graining again. In comparison to the relation $U^*_{t'+t} = U^*_{t'} U^*_t$, we see that the Chapman–Kolmogorov condition can be obtained by demanding that it is allowed to apply a coarse graining, i.e. to reshuffle the phase points within each cell at any intermediate stage of the evolution.

Of course, this coarse graining halfway during the evolution erases all information about the past evolution apart from the label of the cell where the state is located at that time; and this ties in nicely with the view of the Markov property as having no memory (cf. the discussion in Section 10.2).

What is more, the *repeated* application of the coarse graining does lead to a monotonic non-decrease of the Gibbs entropy: If, for simplicity, we divide a time interval into $m$ segments of duration $\tau$, we have

$$\rho_{m\tau} = \underbrace{\mathcal{CG} U^*_\tau \mathcal{CG} U^*_\tau \cdots \mathcal{CG} U^*_\tau}_{m \text{ times}} \rho \tag{10.40}$$

and, since $\sigma(\mathcal{CG}\rho) \geq \sigma(\rho)$ :

$$\sigma[\rho_{m\tau}] \geq \sigma[\rho_{(m-1)\tau}] \geq \ldots \geq \sigma[\rho_\tau] \geq \sigma[\rho_0] \tag{10.41}$$

But since the choice of $\tau$ is arbitrary, we may conclude that $\sigma[\rho_t]$ is monotonically non-decreasing.

Thus, van Kampen argues, the ingredient to be added to the dynamical evolution is that, at any stage of the evolution, one should apply a coarse graining of the distribution. It is important to note that it is not sufficient to do that just once at a single instant. At every stage of the evolution we need to coarse grain the distribution again and again. Van Kampen (1962: 193) calls this the *repeated randomness* assumption.

What is the justification for this assumption? Van Kampen points out that it is 'not unreasonable' (van Kampen, 1962: 182), because of the brute fact of its success in phenomenological physics. Thermodynamics and other phenomenological descriptions of macroscopic systems (the diffusion equation, transport equations, hydrodynamics, the Fokker–Planck equation, etc.) all characterize macroscopic systems with a very small number of variables. This means that their state descriptions are very coarse in comparison with the microscopic phase space. But their evolution equations are autonomous and deterministic: the change of the macroscopic variables is given in terms of the instantaneous values of those very same variables. The success of these equations shows, apparently, that the precise microscopic state does not add any relevant information beyond this coarse description. At the same time, van Kampen admits that the coarse-graining procedure is clearly not always successful. It is not difficult to construct a partition of a phase space into cells for which the Markov property fails completely.

Apparently, the choice of the cells must be 'just right' (van Kampen, 1962: 183). But there is as yet no clear prescription how this is to be done. Van Kampen (1981: 80) argues that it is 'the art of the physicist' to find the right choice, an art in which he or she succeeds in practice by a mixture of general principles and ingenuity, but where no general guidelines can be provided. The justification of the repeated randomness assumption is that it leads to the Markov property and from there onwards to the master equation, providing a successful autonomous, deterministic description of the evolution of the coarse-grained distribution.

It is worth noting that van Kampen thus qualifies the 'usual' point of view on the choice of the cells; namely, that the cells are chosen in correspondence to our finite observation capabilities. Observability of the macroscopic variables is not sufficient for the success of the repeated randomness assumption. It is conceivable (and occurs in practice) that a particular partition in terms of observable quantities does not lead to a Markov process. In that case, the choice of observable variables is simply inadequate and has to be extended with other (unobservable) quantities until we (hopefully) obtain an exhaustive set, i.e. a set of variables for which the evolution can be described autonomously. An example is the spin-echo experiment: the (observable) total magnetization of the system does not provide a suitable coarse-grained description. For further discussion of this theme, see: (Blatt, 1959; Ridderbos and Redhead, 1998; Lavis, 2004; Balian, 2005).

Apart from the unsolved problem for which partition the repeated randomness assumption is to be applied, other objections have been raised against the repeated randomness assumption. Van Kampen actually gives us not much more than the advice to accept the repeated randomness assumption bravely, not to be distracted by its dubious status, and firmly keep our eyes on its success. For authors as Sklar (1993) this puts the problem on its head. They request a justification of the assumption that would *explain* the success of the approach. (Indeed, even van Kampen (1981: 80) describes this success as a 'miraculous fact'!) Such a request, of course, will not be satisfied by a justification that relies on its success. (But that does not mean, in my opinion, that it is an invalid form of justification.)

Another point that seems repugnant to many authors, is that the repeated coarse-graining operations appear to be added 'by hand', in deviation from the true dynamical evolution provided by $U_t$. The increase of entropy and the approach to equilibrium would thus apparently be a consequence of the fact that *we* shake up the probability distribution repeatedly in order to wash away all information about the past, while refusing a dynamical explanation for this procedure. Redhead (1995: 31) describes this procedure as 'one of the most deceitful artifices I have ever come across in theoretical physics' (see also Blatt, 1959; Sklar, 1993; Callender, 1999, for similar objections).

### *10.5.2 Interventionism or 'open systems'*

Another approach to stochastic dynamics is by reference to open systems. The idea is here that the system is in continual interaction with the environment, and that this is responsible for the approach to equilibrium.

Indeed, it cannot be denied that in concrete systems isolation is an unrealistic idealization. The actual effect of interaction with the environment on the microscopic evolution can be enormous. A proverbial example, going back to Borel (1914), estimates the gravitational effect caused by displacing one gram of matter on Sirius by one centimetre on the microscopic evolution of an earthly cylinder of gas. Under normal conditions, the effect is so large, that, roughly and for a typical molecule in the gas, it may be decisive for whether or not this molecule will hit another given molecule after about 50 intermediary collisions. That is to say: microscopic dynamical evolutions corresponding to the displaced and the undisplaced matter on Sirius start to diverge considerably after a time of about $10^{-6}$ s. In other words, the mechanical evolution of such a system is so extremely sensitive for disturbances of the initial state that even the most minute changes in the state of the environment can be responsible for large changes in the microscopic trajectory. But we cannot control the state of environment. Is it possible to regard irreversible behaviour as the result of such uncontrollable disturbances from outside?[6]

---

[6] Note that the term 'open system' is employed here for a system in (weak) interaction with its environment. This should be distinguished from the notion of 'open system' in other branches of physics where it denotes a system that can exchange particles with its environment.

Let $(x, y)$ be the state of a total system, where, as before, $x \in \Gamma^{(s)}$ represents the state of the object system and $y \in \Gamma^{(e)}$ that of the environment. We assume that the total system is governed by a Hamiltonian of the form

$$H_{\text{tot}}(x, y) = H_{(s)} + H_{(e)} + \lambda H_{\text{int}}(x, y) \qquad (10.42)$$

so that the probability density of the ensemble of total systems evolves as

$$\rho_t(x, y) = U_t^* \rho_0(x, y) = \rho(U_{-t}(x, y)) \qquad (10.43)$$

i.e. a time-symmetric, deterministic and measure-preserving evolution.

At each time, we may define marginal distributions for both system and environment:

$$\rho_t^{(s)}(x) = \int dy\, \rho_t(x, y) \qquad (10.44)$$

$$\rho_t^{(e)}(x) = \int dx\, \rho_t(x, y) \qquad (10.45)$$

We are, of course, mostly interested in the object system, i.e. in (10.44). Assume further that at time $t = 0$ the total density factorizes:

$$\rho_0(x, y) = \rho_0^{(s)}(x) \rho_0^{(e)}(y) \qquad (10.46)$$

What can we say about the evolution of $\rho_t^{(s)}(x)$? Does it form a Markov process, and does it show increase of entropy?

An immediate result (see e.g. Penrose and Percival, 1962) is this:

$$\sigma[\rho_t^{(s)}] + \sigma[\rho_t^{(e)}] \geq \sigma[\rho_0^{(s)}] + \sigma[\rho_0^{(e)}] \qquad (10.47)$$

where $\sigma$ denotes the Gibbs fine-grained entropy

$$\sigma[\rho] = -\int \rho(x) \ln \rho(x)\, dx \qquad (10.48)$$

This result follows from the fact that $\sigma[\rho_t]$ is conserved and that the entropy of a joint probability distribution is always smaller than or equal to the sum of the entropies of their marginals; with equality if the joint distribution factorizes. This gives a form of entropy change for the total system, but it is not sufficient to conclude that the object system itself will evolve towards equilibrium, or even that its entropy will be monotonically increasing. (Note that (10.47) holds for $t \leq 0$ too.)

Actually, this is obviously not to be expected. There are interactions with an environment that may lead the system away from equilibrium. We shall have to make additional assumptions about the situation. A more or less usual set of assumptions is:

(a) The environment is very large (or even infinite); i.e. the dimension of $\Gamma^{(e)}$ is much larger than that of $\Gamma^{(s)}$, and $H_{(s)} \ll H_{(e)}$.
(b) The coupling between the system and the environment is weak, i.e. $\lambda$ is very small.
(c) The environment is initially in thermal equilibrium, e.g. $\rho^{(e)}(y)$ is canonical:

$$\rho_0^{(e)} = \frac{1}{Z(\beta)} e^{-\beta H^{(e)}} \tag{10.49}$$

(d) One considers time scales only that are long with respect to the relaxation times of the environment, but short with respect to the Poincaré recurrence time of the total system.

Even then, it is a major task to obtain a master equation for the evolution of the marginal state (10.44) of the system, or to show that its evolution is generated by a semigroup, which would guarantee that this forms a Markov process (under the proviso of footnote 4). Many specific models have been studied in detail (cf. Spohn, 1980). General theorems were obtained (although mostly in a quantum mechanical setting) by (Davies, 1974; 1976a; Gorini *et al.*, 1976; Lindblad, 1976). But there is a similarity to the earlier approach: it seems that, here too, an analogue of 'repeated randomness' must be introduced (Mehra and Sudarshan, 1972; van Kampen, 1994; Maes and Netočný, 2003).

At the risk of oversimplifying the results obtained in this analysis, I believe they can be summarized as showing that, in the so-called 'weak coupling' limit, or some similar limiting procedure, the time development of (10.44) can be modelled as

$$\rho_t^{(s)}(x) = T_t \rho^{(s)}(x) \, t \geq 0 \tag{10.50}$$

where the operators $T_t$ form a semigroup, while the environment remains in its steady equilibrium state:

$$\rho_t^{(e)}(y) = \rho_0^{(e)}(y) \, t \geq 0 \tag{10.51}$$

The establishment of these results would also allow one to infer, from (10.47), the monotonic non-decrease of entropy of the system.

To assess these findings, it is convenient to define, for a fixed choice of $\rho_0^{(e)}$ the following linear map on probability distributions of the total system:

$$\mathcal{TR} : \rho(x, y) \mapsto \mathcal{TR}\rho(x, y) = \int \rho(x, y) \, dy \cdot \rho_0^{(e)}(y) \tag{10.52}$$

This map removes the correlation between the system and the environment, and projects the marginal distribution of the environment back to its original equilibrium form.

Now, it is not difficult to see that the Chapman–Kolmogorov equation (which is equivalent to the semigroup property) can be expressed as

$$\mathcal{TRU}^*_{t+t'} = \mathcal{TRU}^*_{t'}\mathcal{TRU}^*_t \qquad \text{for all } t, t' \geq 0 \tag{10.53}$$

which is analogous to (10.39).

There is thus a strong formal analogy between the coarse-graining and the open-systems approaches. Indeed, the variables of the environment play a role comparable to the internal coordinates of a cell in the coarse-graining approach. The exact microscopic information about the past is here translated into the form of correlations with the environment. This information is now removed by assuming that at later times, effectively, the state may be replaced by a product of the form (10.46), neglecting the back-action of the system on the environment. The mappings $\mathcal{CG}$ and $\mathcal{TR}$ are both linear and idempotent mappings, that can be regarded as special cases of the projection operator techniques of Nakajima and Zwanzig, which allows for a more systematical and abstract elaboration, sometimes called *subdynamics*.

Some proponents of the open-systems approach, (e.g. Morrison, 1966; Redhead, 1995), argue that in contrast to the coarse-graining approach, the present procedure is 'objective'. Presumably, this means that there is supposed to be a fact of the matter about whether the correlations are indeed 'exported to the environment'. However, the analogy between both approaches makes one suspect that any problem for the coarse-graining approach is translated into an analogous problem of the open-systems approach. Indeed, the problem of finding a privileged partition that we discussed in the previous section is mirrored here by the question where one should place the division between the 'system' and 'environment'. There is no doubt that in practical applications this choice is also arbitrary.

## 10.6 Can the Markov property explain irreversible behaviour?

### Ad (v)

Finally, I turn to what may well be the most controversial and surprising issue: is the Markov property, or the repeated randomness assumption offered to motivate it, responsible for the derivation of time-reversal non-invariant results?

We have seen that every non-invertible homogeneous Markov process displays 'irreversible behaviour' in the sense that different initial probability distributions will tend to become more alike in the course of time. Under certain technical conditions, one can obtain stronger results, e.g. an approach to a unique equilibrium state, monotonic non-decrease of absolute entropy, etc. All these results seem to be clearly time-asymmetric. And yet we have also seen that the Markov property is explicitly time-symmetric. How can these be reconciled?

To start off, it may be noted that it has often been affirmed that the Markov property is the key towards obtaining time-asymmetric results. For example, Penrose writes:

> ... the behaviour of systems that are far from equilibrium is not symmetric under time reversal: for example: heat always flows from a hotter to a colder body, never from a colder to a hotter. If this behaviour could be derived from the symmetric laws of dynamics alone there would, indeed, be a paradox; we must therefore acknowledge the fact that some additional postulate, non-dynamical in character and asymmetric under time reversal must be adjoined to the symmetric laws of dynamics before the theory can become rich enough to explain non-equilibrium behaviour. In the present theory, this additional postulate is the Markov postulate.
>
> *(Penrose, 1970: 41)*

Many similar statements, e.g. that the repeated randomness assumption is 'the additional element by which statistical mechanics has to be supplemented in order to obtain irreversible equations' (van Kampen, 1962: 182), or that the non-invertibility of a Markov process provides the origin of thermodynamic behaviour (Mackey, 1992) can be found in the works of advocates of this approach.

But how can this be, given that the Markov property is explicitly time-symmetric? In order to probe this problem, consider another question. How does a given probability distribution $P(y, 0)$ evolve for negative times? So, starting again from (10.6), let us now take $t \leq 0$. We still have:

$$P(y, t) = \sum_{y'} P(y, t, | y', 0) P(y', 0) \tag{10.54}$$

These conditional probabilities $P(y, t, | y', 0)$ satisfy the 'time-reversed' Markov property (10.10), that says that extra specification of later values is irrelevant for the retrodiction of earlier values. As a consequence, we get for $t \leq t' \leq t'', 0$:

$$P(y, t | y'', t'') = \sum_{y'} P(y, t | y', t') P(y', t' | y'', t'') \tag{10.55}$$

i.e. a time-reversed analogue of the Chapman–Kolmogorov equation.

We may thus also consider these conditional probabilities for negative times as backward evolution operators. If we could assume their invariance under time translation, i.e. that they depend only on the difference $\tau = t - t'$, we could write

$$S_\tau(y | y') := P(y, t | y, t') \text{ with } \tau = t - t' \leq 0 \tag{10.56}$$

and obtain a second semigroup of operators $S_\tau$, obeying

$$S_{\tau + \tau'} = S_\tau \circ S_{\tau'} \tau, \tau' \leq 0 \tag{10.57}$$

that generate stochastic evolutions towards the past.

These backward conditional probabilities are connected to the forward conditional probabilities by means of Bayes' theorem:

$$P_{(1|1)}(y, t|y', t') = \frac{P_{(1|1)}(y', t'|y, t)P(y, t)}{P(y', t')} \tag{10.58}$$

and if the process, as before, is homogeneous this becomes

$$P_{(1|1)}(y, t|y', t') = \frac{T_{-\tau}(y'|y)P_t(y)}{P_{t'}(y')}; \tau = t - t' < 0 \tag{10.59}$$

The matrix $P_{(1|1)}(y, t|y', t')$ always gives for $t < t'$ the correct 'inversion' of $T_t$. That is to say:

$$\sum_{y'} P(y, t|y', t')(T_{t'-t}P_t)(y') = P_t(y) \tag{10.60}$$

Note firstly that (10.59) is *not* the matrix-inverse of $T_t$! Indeed, the right-hand side of (10.59) depends on $P_t$ and $P_{t'}$ as well as $T$. Even if the matrix-inverse $T^{(inv)}$ does not exist, or is not a bona fide stochastic matrix, the evolution towards the past is governed by the Bayesian inversion, i.e. by the transition probabilities (10.59).

Note also that if the forward transition probabilities are homogeneous, this is not necessarily so for the backward transition probabilities. For example, if in (10.59) one translates both $t$ and $t'$ by $\delta$, one finds

$$P(y, t+\delta|y', t'+\delta) = \frac{T_{-\tau}(y'|y)P(y, t+\delta)}{P(y', t'+\delta)}$$

Here, the right-hand side generally still depends on $\delta$. In the special case that the initial distribution is itself stationary, the backward transition probabilities are homogeneous whenever the forward ones are. If $P(y, t)$ is not stationary, we might still reach the same conclusion, as long as the non-stationarity is restricted to those elements $y$ or $y'$ of $\mathcal{Y}$ for which $T_t(y|y') = 0$ for all $t$. Otherwise, the two notions become logically independent.

This gives rise to an unexpected new problem. Usually, an assumption of homogeneity (or time translation invariance) is seen as philosophically innocuous, as compared to time-reversal invariance. But here we see that assuming time-translation invariance for a system of *forward* transition probabilities is not equivalent to assuming the same invariance for the *backward* transition probabilities. If one believes that one of the two is obvious, how will one go about explaining the failure of the other? And how would one explain the preference for which one of the two is obvious, without falling into the 'double standards' accusation of the kind raised by Price (1996)?

But what about entropy increase? We have seen before that for every non-invertible Markov process the relative entropy of the distribution $P$ with respect to the equilibrium distribution increases, and that the distribution evolves towards equilibrium. (Homogeneity of the process is not needed for this conclusion.) But the backward evolution operators form a Markov process too, for which exactly the same holds. This seems paradoxical. If $T_t P_0 = P_t$, we also have $P_t = S_{-t} P_0$. The entropy of $P_t$ can hardly be both higher and lower than that of $P_0$! An example may clarify the resolution of this apparent problem: namely, the stationary solutions of $S$ are not the same as the stationary solutions of $T$!

*Example* Consider a Markov chain with $\mathcal{Y} = \{1, 2\}$ and let

$$T = \begin{pmatrix} \frac{1}{2} & \frac{1}{2} \\ \frac{1}{2} & \frac{1}{2} \end{pmatrix}. \tag{10.61}$$

Choose an initial distribution $P_0 = (\alpha, 1 - \alpha)$. After one step we already get: $P_1 = T P_0 = \left(\frac{1}{2}, \frac{1}{2}\right)$ which is also the (unique) stationary distribution $P^*$. The backward transition probabilities are given by Bayes' theorem, and one finds easily:

$$S = \begin{pmatrix} \alpha & \alpha \\ 1 - \alpha & 1 - \alpha \end{pmatrix} \tag{10.62}$$

The stationary distribution for this transition probability is $\tilde{P}^* = (\alpha, 1 - \alpha)$. That is to say: for the forward evolution operator the transition

$$\begin{pmatrix} \alpha \\ 1 - \alpha \end{pmatrix} \xrightarrow{T} \begin{pmatrix} \frac{1}{2} \\ \frac{1}{2} \end{pmatrix} \tag{10.63}$$

is one for which a non-stationary initial distribution evolves towards a stationary one. The relative entropy increases: $H(P_0, P^*) \leq H(P_1, P^*)$. But for the backward evolution, similarly:

$$\begin{pmatrix} \frac{1}{2} \\ \frac{1}{2} \end{pmatrix} \xrightarrow{S} \begin{pmatrix} \alpha \\ 1 - \alpha \end{pmatrix} \tag{10.64}$$

represents an evolution from a non-stationary initial distribution to the stationary distribution $\tilde{P}^*$ and, here too, relative entropy increases: $H(P_1, \tilde{P}^*) \leq H(P_0, \tilde{P}^*)$.

The illusion that non-invertible Markov processes possess a built-in time-asymmetry is (at least partly) due to the habit of regarding $T_\tau$ as a fixed evolution operator on an independently chosen distribution $P_0$. Such a view is of course very familiar in other problems in physics, where deterministic evolution operators generally *do* form a group and may be used, at our heart's desire, for positive and negative times.

Indeed, the fact that these operators in general have no inverse might seem to reflect the idea that Markov processes have no memory and 'lose information' along the way and that is the cause of the irreversible behaviour, embodied in the time-asymmetric master equation, increase of relative or absolute entropy or approach to equilibrium. But actually, every Markov process has apart from a system of forward, also a system of backward transition probabilities, that again forms a semigroup (when they are homogeneous). If we had considered *them* as given we would get all conclusions we obtained before, but now for negative times.

I conclude that irreversible behaviour is not built into the Markov property, or in the non-invertibility of the transition probabilities (or in the repeated randomness assumption,[7] or in the master equation or in the semigroup property). Rather the appearance of irreversible behaviour is due to the choice to rely on the forward transition probabilities, and not the backward. A similar conclusion has been reached before by Edens (2001) in the context of proposals of Prigogine and his coworkers. My main point here is that the same verdict also holds for more 'mainstream' approaches as coarse graining or open systems.

## 10.7 Reversibility of stochastic processes

In order not to end this chapter on a destructive note, let me emphasize that I do not claim that the derivation of irreversible behaviour in stochastic dynamics is impossible. Instead, the claim is that motivations for desirable properties of the forward transition probabilities are not enough; one ought also show that these properties are lacking for the backward transitions.

In order to set up the problem of irreversibility in this approach to non-equilibrium statistical mechanics for a more systematic discussion, one first ought to provide a reasonable definition for what it means for a stochastic process to be (ir)reversible; a definition that would capture the intuitions behind its original background in Hamiltonian statistical mechanics.

One general definition that seems to be common (cf. Kelly, 1979: 5) is to call a stochastic process reversible iff, for all $n$ and $t_1, \ldots, t_n$ and $\tau$:

$$P_{(n)}(y_1, t_1; \ldots; y_n, t_n) = P_{(n)}(y_1, \tau - t_n; \ldots; y_n, \tau - t_n) \qquad (10.65)$$

See Grimmett and Stirzaker (1982: 219) for a similar definition restricted to Markov processes. The immediate consequence of this definition is that a stochastic process

---

[7] In recent work, van Kampen acknowledges that the repeated randomness assumption by itself does not lead to irreversibility: 'This repeated randomness assumption [...] breaks the time symmetry by explicitly postulating the randomization *at the beginning* of the time interval $\Delta t$. There is no logical justification for this assumption other than that it is the only thing one can do and that it works. If one assumes randomness at the end of each $\Delta t$ coefficients for diffusion, viscosity, etc. appear with the wrong sign; if one assumes randomness at the midpoint no irreversibility appears' (van Kampen 2002: 475, original emphasis).

can only be reversible if the single-time probability $P_{(1)}(y, t)$ is stationary, i.e. in statistical equilibrium. Indeed, this definition seems to make the whole problem of reconciling irreversible behaviour with reversibility disappear. As Kelly (1979: 19) notes in a discussion of the Ehrenfest model: 'there is no conflict between reversibility and the phenomenon of increasing entropy – reversibility is a property of the model in equilibrium and increasing entropy is a property of the approach to equilibrium'.

But clearly, this view trivializes the problem, and therefore it is not the appropriate definition for non-equilibrium statistical mechanics. Recall that the Ehrenfest dog flea model (Section 10.2) was originally proposed in an attempt of showing how a tendency of approaching equilibrium from an initial non-equilibrium distribution (e.g. a probability distribution that gives probability 1 to the state that all fleas are located on the same dog) could be reconciled with a stochastic yet time-symmetric dynamics.

If one wants to separate considerations about initial conditions from dynamical considerations at all, one would like to provide a notion of (ir)reversibility that is associated with the stochastic dynamics alone, independent of whether the initial distribution is stationary.

It seems that an alternative definition which would fulfil this intuition is to say that a stochastic process is reversible if, for all $y$ and $y'$ and $t' > t$,

$$P_{(1|1)}(y, t|y', t') = P_{(1|1)}(y, t'|y', t). \tag{10.66}$$

In this case we cannot conclude that the process must be stationary, and indeed, the Ehrenfest model would be an example of a reversible stochastic process. I believe this definition captures the intuition that if at some time state $y'$ obtains, the conditional probability of the state one time-step earlier being $y$ is equal to that of the state one time-step later being $y$. A similar proposal has been advocated by Bacciagaluppi (2007).

According to this proposal, the aim of finding the 'origin' of irreversible behaviour or 'time's arrow', etc. in stochastic dynamics must then lie in finding and motivating conditions under which the forward transition probabilities are different from the backwards transition probabilities, in the sense of a violation of (10.66). Otherwise, irreversible behaviour would essentially be a consequence of the assumptions about initial conditions, a result that would not be different in principle from conclusions obtainable from Hamiltonian dynamics.

## References

Bacciagaluppi, G. (2007). Probability and time symmetry in classical Markov Processes. http://philsci-archive.pitt.edu/archive/00003534/

Balian, R. (2005). Information in statistical physics. *Studies In History and Philosophy of Modern Physics*, **36**, 323–353.

Blatt, J. M. (1959). An alternative approach to the ergodic problem. *Progress in Theoretical Physics*, **22**, 745–756.

Borel, E. (1914). *Le Hasard*. Paris: Alcan.

Callender, C. (1999). Reducing thermodynamics to statistical mechanics: the case of entropy. *Journal of Philosophy*, **96**, 348–373.

Davies, E. B. (1974). Markovian master equations. *Communications in Mathematical Physics*, **39**, 91–110.

Davies, E. B. (1976a). Markovian master equations II. *Mathematische Annalen*, **219**, 147–158.

Davies, E. B (1976b). *Quantum Theory of Open Systems*. New York: Academic Press.

Edens, B. (2001). Semigroups and symmetry: an investigation of Prigogine's theories. http://philsci-archive.pitt.edu/archive/00000436/.

Ehrenfest, P. and Ehrenfest, T. (1907). Über Zwei Bekannte Einwände gegen das Boltzmannsche $H$-Theorem. *Phyikalische Zeitschrift*, **8**, 311–314.

Gantmacher, F. R. (1959). *Matrizenrechnung*, vol. 2. Berlin: Deutscher Verlag der Wissenschaften.

Gorini, V., Kossakowski, A. and Sudarshan, E. C. G. Completely positive dynamical semigroups of $N$-level systems. *Journal of Mathematical Physics*, **17**, 8721–8825.

Grimmett, G. R. and Stirzaker, D. R. (1982). *Probability and Random Processes*. Oxford: Clarendon Press.

van Kampen, N. G. (1962). Fundamental problems in the statistical mechanics of irreversible processes. In *Fundamental Problems in Statistical Mechanics*, ed. E. G. D. Cohen. Amsterdam: North-Holland, pp. 173–202.

van Kampen, N. G. (1981). *Stochastic Processes in Chemistry and Physics*. Amsterdam: North-Holland.

van Kampen, N. G. (1994) Models for dissipation in quantum mechanics. In *25 Years of Non-equilibrium Statistical Mechanics*, ed. J. J. Brey *et al.* Berlin: Springer-Verlag.

van Kampen, N. G. (2002) The road from molecules to Onsager. *Journal of Statistical Physics*, **109**, 471–481.

Kelly F. P. (1979). *Reversibility and Stochastic Networks*. Chichester: Wiley. Also at http://www.statslab.cam.ac.uk/ afrb2/kelly_book.html.

Lavis, D. A. (2004). The spin-echo system reconsidered. *Foundations of Physics*, **34**, 669–688.

Lindblad, G. (1976). On the generators of quantum dynamical semigroups. *Communications in Mathematical Physics*, **48**, 119–130.

Lindblad, G. (1983). *Non-equilibrium Entropy and Irreversibility*. Dordrecht: Reidel.

Maes, C. and Netočný, K. (2003). Time-reversal and entropy. *Journal of Statistical Physics*, **110**, 269–310.

Mackey, M. C. (1992). *Time's Arrow: the Origins of Thermodynamic Behavior*. New York: Springer-Verlag.

Mackey, M. C. (2001). Microscopic dynamics and the second law of thermodynamics. In *Time's Arrows, Quantum Measurements and Superluminal Behavior*, ed. C. Mugnai, A. Ranfagni and L. S. Schulman. Rome: Consiglio Nazionale delle Ricerche.

Mehra, J. and Sudarshan, E. C. G. (1972). Some reflections on the nature of entropy, irreversibility and the second law of thermodynamics. *Nuovo Cimento B*, **11**, 215–256.

Moran, P. A. P. (1961). Entropy, Markov processes and Boltzmann's H-theorem. *Proceedings of the Cambridge Philosophical Society*, **57**, 833–842.

Morrison, P. (1966). Time's arrow and external perturbations. In *Preludes in Theoretical Physics in Honor of V. F. Weisskopf*, A. de Shalit, H. Feshbach and L. van Hove. Amsterdam: North-Holland, pp. 347–351.

Penrose, O. (1970). *Foundations of Statistical Mechanics: a Deductive Treatment*. Oxford: Pergamon Press.

Penrose, O. and Percival, I. (1962). The direction of time. *Proceedings of the Physical Society*, **79**, 605–616.

Petersen, K. (1983). *Ergodic Theory*. Cambridge: Cambridge University Press.

Price, H. (1996). *Time's Arrow and Archimedes' Point*. New York: Oxford University Press.

Redhead, M. (1995). *From Physics to Metaphysics*. Cambridge: Cambridge University Press.

Ridderbos, T. M. (2002). The coarse-graining approach to statistical mechanics: how blissful is our ignorance. *Studies in History and Philosophy of Modern Physics*, **33**, 65–77.

Ridderbos, T. M and Redhead, M. L. G. (1998). The spin-echo experiment and the second law of thermodynamics. *Foundations of Physics*, **28**, 1237–1270.

Sklar, L. (1993). *Physics and Chance. Philosophical Issues in the Foundations of Statistical Mechanics*. Cambridge: Cambridge University Press.

Spohn, H. (1980). Kinetic equations from Hamiltonian dynamics: Markovian limits. *Reviews of Modern Physics*, **52**, 569–615.

Streater, R. F. (1995). *Statistical Dynamics; a Stochastic Approach to Non-equilibrium Thermodynamics*. London: Imperial College Press.

Sudarshan, E. C. G., Mathews, P. M. and Rau, J. (1961). Stochastic dynamics of quantum-mechanical systems. *Physical Review*, **121**, 920–924.

Uffink, J. (2007) Compendium of the foundations of classical statistical physics. In *Handbook of the Philosophy of Physics,* ed. J. Butterfield and J. Earman. Amsterdam: North-Holland, pp. 923–1074.

# Index

Abraham, Ralph 93
Albert, David 2, 4, 5, 6, 13–14, 15, 20, 25, 26–29, 32, 35, 55, 100, 101, 103, 106, 109, 111, 112, 115–116
Anjum, R. L. 126
Armstrong, David 119, 124
Arnold, Vladimir 99
arrows of time 3–4, 59–62, 66, 205
   cosmological arrow of time 67
   gravitational arrow of time 62
   master arrow of time 59, 67
   quantum-mechanical arrow of time 61, 66
   radiation arrow of time 59–60, 66
   thermodynamic arrow of time 21, 26, 34, 61
asymmetry 34
   asymmetry of decision counterfactuals 14, 28–32
   asymmetry of radiation 13
   asymmetry of time *see* time
   Boltzmannian account of the thermodynamic asymmetry *see* Boltzmann
   causal asymmetry 3, 4, 13, 14
Avez, A. 99

Baez, J. 20
Balescu, R. 53
Balian, R. 196
Balzer, Wolfgang 139, 142
Banach space 188
Bartels, A. 131
Batterman, Robert 7, 8, 166, 170
Bayes' theorem 202, 203
   Bayesian inversion 14, 202
   Bayesianism 115
Beebee, Helen 119, 120, 122
Bennett, Jonathan 23–24
Bertrand's paradoxes 73, 81
big bang 15, 25, 35, 62, 63, 65, 66, 94, 101, 120, 123, 126
Binney, J. J. 45, 50
Biot, Jean-Baptiste 142
Bird, Alexander 119, 122, 125, 131
Black, R. 129

black holes 62–63
   black hole entropy *see* entropy
Blatt, J. M. 181, 196, 197
Bleher, P. M. 168
Bojowald, M. 65
Boltzmann, Ludwig 1, 6, 34–36, 38, 48, 50, 61, 92, 95–96, 97
   Boltzmann entropy *see* entropy
   Boltzmann equation 39, 48–49, 53, 183, 189
   Boltzmannian account of the thermodynamic asymmetry 4, 13, 15, 16, 24, 26, 56; *see also* past-hypothesis
   Boltzmannian probability 46, 95
   Boltzmannian statistical mechanics *see* statistical mechanics
   Boltzmannian version of the second law of thermodynamics 94, 101
Borel, Émile 197

Caamaño, M. C. 142
Callen, H. B. 142
Callender, Craig 4, 35, 56, 94, 101, 124, 182, 197
caloric theory 8, 142, 145
Carnap, Rudolf 83
Carnot, Nicolas 142, 153
   Carnot principle 143, 146
Cassandro, M. 168, 173
causal handles 14, 26–28, 29
causal theory of properties 126–133
causation 2, 3, 14, 121, 126, 130
   asymmetry of causation *see* asymmetry
   backward causation 14, 202
central limit theorem 173
chance *see* probability
Chapman–Kolmogorov equation 185–188, 195, 200, 201
Chavanis, P. H. 51, 53
Choi, S. 122
Clapeyron, Benoît 142, 153
   Clapeyron's law 155
Clark, Peter 95
Clausius, Rudolf J. 7, 139–157
coarse-graining 8, 23, 25, 61, 180, 194–205

Corry, R. 130
cosmology 22, 25, 35, 40, 61
  cosmological arrow of time *see* arrows of time
Cross, T. 122

Dabrowski, M. P. 65
Dauxois, T. 43
Davies, E. B. 181, 199
DeRoeck, W. 39, 52, 54
determinism 1–2, 16, 29, 77, 79, 95, 103, 122, 123, 126, 180
  Laplacean determinism 122
  quasi-determinism 18, 20, 21–22
Dorato, M. 131
Dunning-Davies, J. 44
dynamics 4, 93, 99
  dynamical laws 6, 16, 18, 24, 40, 42, 92, 100
  dynamical system 16
  galactic dynamics, stellar dynamics 52–53
  stochastic dynamics 8, 180–205

Eagle, E. 71
Earman, John 16, 22, 36, 42, 47, 95, 99, 101
Eddington, Arthur 59
Edens, B. 204
Ehrenfest, Paul and Tatiana 6, 95, 98, 101, 183
  Ehrenfest dog flea model 183, 205
Einstein, Albert 35
  Einstein equations 36, 62, 63, 66
electromagnetism 40
Elga, Adam 107
Elkana, Y. 142
Ellis, Brian 119, 125
emergence 7, 8, 59, 61, 142, 169, 171
empiricism 119
energy
  conservation of energy in a Hamiltonian system 93
  interaction energy 46
  total energy 44, 49
Engel, E. 85
ensembles
  equivalence and inequivalence of ensembles 178
entropy
  Bekenstein entropy 36, 62
  black hole entropy 42
  Boltzmann entropy 4, 34–37, 38–39, 40, 42, 46, 47, 49, 52–56, 93–94, 115, 122
  entropy of the universe 25, 62
  Gibbs entropy 47, 54, 115, 182, 195, 198
  increase of entropy 189–193
  principle of maximum entropy 115
  Shannon entropy 115
  thermodynamic entropy 34, 171, 178
  Tsallis entropy 47
equilibrium
  approach to equilibrium 97, 113, 183, 189–193, 205

ergodicity 6, 42, 95, 99–100, 192
Esfeld, Michael 6, 130, 131, 132, 133
Exner, Franz 1

Fedele, J. B. 53
Fermat, Pierre de 73
Feynman, Richard 35
Field, H. H.
Fine, Kit 21
Fokker–Planck equation 53–54, 55, 196
Frank, T. 54
free will 122
Friedman–Robertson–Walker metric 36
Frigg, Roman 6, 39, 95, 123
Frisch, Mathias 3, 13, 19, 28
Frobenius, Ferdinand Georg 191

Gallavotti, Giovanni 175
Galton board 20–21, 28, 85
Gantmacher, F. R. 191
Garrido, P. L. 49
gases
  ideal gas law 4, 59, 145, 149, 153, 154, 156
  kinetic theory of gases 143
Gaussian distribution 167, 169, 171, 176
general relativity theory 2, 34, 36, 47, 62, 63
  quantum general relativity 63
Ghirardi, G. 131
Gibbs, J. Willard 8, 38, 142, 159–160, 161, 162, 175
  Gibbs entropy *see* entropy
  Gibbsian thermodynamics 143
Goldenfeld, Nigel 170
Goldstein, Herbert 93
Goldstein, Sheldon 39, 49, 115
Gorini, V. 199
gravity 4, 34–56, 64
  gravitational arrow of time *see* arrows of time
  quantum gravity 4, 42, 59–67
Green, M. S. 53
Grimmet, G. R. 204
Gunderson, L. 122

Hamiltonian mechanics 109, 111, 112, 113, 204
  Hamilton's equations 38, 39, 43, 66, 93, 101, 181
Handfield, T. 126
Hawking effect 62
Hawking–Page measure 36
heat
  Clausius' mechanical theory of heat 143
  heat death 60
  heat differential 148
  principle of the equivalence of heat and work 143, 151
Heggie, D. C. 43, 53
Heidelberger, M. 73, 83
Heil, John 119, 125
von Helmholtz, Hermann 143
Hertel, P. 46
Hitchcock, Christopher 2, 101
Hoefer, Carl 95, 115

Holland–Wald measure 36
Holton, R. 29, 30
Hopf, Eberhard 85
*H*-theorem 48, 49, 51–52, 53–54, 99, 181
Hume, David 120
　　Humean chance 102, 115
　　Humean metaphysics *see* metaphysics
　　Humean regularity view of causation 121
　　Humean supervenience 119, 121
Hut, P. 43, 53

indeterminism 80
induction 122
initial conditions 77, 80, 81, 82, 87, 110, 114, 122, 150, 205
　　initial conditions and boundary conditions 74
interventionism 181, 197–200
　　interventionist approach 193
irreversibility 60, 67
　　irreversibility in stochastic dynamics 180
　　irreversible processes in statistical mechanics 8
　　reversible and irreversible cycles 8, 154, 155–156
isothermal sphere 50

Jayne, Edwin Thompson 115
Jona-Lasinio, G. 168, 173
Joos, E. 61
Joule, James Prescott 143, 144

Kadanoff, Leo 168, 171
Kamenshchik, A. Y. 65
Kandrup, H. 52
Kelly, F. P. 186, 204, 205
Kelvin 143
Keynes, John Maynard 83
Khinchin, Alexander 114, 163, 167, 171
Kiefer, Claus 4, 62, 63, 65, 66, 67
kinetic equation 4, 51–52, 53–56
Kries, Johannes von 73, 80, 82, 83, 84, 87
Kronecker delta 187
Kuhn, Thomas S. 141, 142, 163

Ladyman, J. 130
Lakatos, Imre 141
Lam, V. 131
Landau, Lew Dawidowitsch 43, 44
　　Landau equation 53
Laplace, Pierre-Simon 73, 142
　　Laplacean determinism 122
Lavis, David 38, 94, 99, 196
Lavoisier, Antoine Laurent de 142
law of large numbers 71–72, 76–77
Lebesgue measure 38, 42, 47, 74, 76, 93, 105
Lebowitz, J. L. 49, 93
Levy-Leblond, J.-M. 46
Lewis, David 6, 23, 24, 95, 100, 102, 115, 119–121, 122, 123, 132
Liboff, R. L. 53
Lieb, E. H. 46
Lifshitz, E. 43, 44

Lindblad, G. 181, 199
Liouville's theorem 15, 19, 39, 52, 54, 104, 182
Loewer, Barry 4, 6, 13–14, 16–18, 19, 20, 22, 23–24, 26, 28–30, 100–114, 115, 119, 124, 142
Lynden-Bell, D. 50

Mackey, M. C. 181, 182, 189, 191, 193, 201
Maes, C. 39, 199
magnetization 196
Malzkorn, W. 122
Markov process 8, 181–204
Marsden, Jerrold 93
Martin, Charles 119, 122, 125
master arrow of time *see* arrows of time
master equation 9, 61, 181–183, 188–189, 196–204
matter 25, 35, 44, 60
Maxwell, James Clerk 1, 48
　　Maxwell–Boltzmann probability distribution 42, 48, 51
Meacham, Christopher 115
Mehra, J. 180, 199
Mele, A. 122
metaphysics
　　Humean metaphysics 6, 119–124
　　metaphysics of powers 6, 119, 121, 124–128
　　metaphysics of structures
　　metaphysics of universals 119
von Mises, Richard 95
Moran, P. A. P. 189
Morrison, P. 200
Moulines, C. Ulises 7, 139, 141, 142, 143
$\mu$-space 48, 49–50, 96, 183
Mumford, Stephen 119, 125, 126, 129, 130

Nagel, Ernest 7, 8, 160
Nauenberg, M. 47
Netočný, K. 39, 199
Newtonian mechanics 37, 111, 142

Ogorodnikov, K. F. 50
Olbers' paradox 60
open systems *see* interventionism

Padmanabhan, T. 43, 46
Pascal, Blaise 73
past hypothesis 4, 15, 16–20, 22–23, 24, 27, 34–56, 94, 101, 123
　　past hypothesis statistical postulate 102
Pauli, Wolfgang 188
Penrose, Roger 35, 61, 62, 180, 198, 201
Percival, I. 38–39
Perron–Frobenius theorem 192
Perron, Oskar 191
Petersen, K. 192
phase transition 167, 174, 175–178
Plato, Jan von 1, 85, 95, 98, 99
Poincaré, Henri 84
　　Poincaré recurrence time 199
Poisson's law 149
Popper, Karl 82, 127

Price, H. 202
Prigogine, Ilja 204
probability 5, 114
   epistemic interpretation of statistical mechanics probabilities 115–116
   macro and micro probabilities 6, 95, 100
   objectivist interpretation of probability 5–6, 71, 76, 78, 79, 82
   propensity theory of probability 72
   range interpretation of probability 73–90
Psillos, Stathis 130

quantum-mechanical arrow of time *see* arrows of time

radiation arrow of time *see* arrows of time
randomness
   repeated randomness assumption 182, 194–197, 200–201, 204
Rankine, William 143
realism
   empiricist structural realism 132
   ontic structural realism 130–131
Rédei, Miklos 99
Redhead, Michael 115, 182, 196, 197, 200
reduction of thermodynamics to statistical mechanics 3, 7, 159–178
Regnault, Henri Victor 143, 150, 154, 156
Reichenbach, Hans 55
reversibility 40
   reversibility objection 15, 23
   reversibility of stochastic processes 183, 204–205
Ridderbos, T. M. 181, 196
Rimini, A. 131
Risken, H. 54
Rosenthal, Jacob 5
Ross, D. 130
Rovelli, C. 63
Rowlinson, J. S. 43
Russell, Bertrand 130

Saslaw, W. C. 43
Schrödinger, Erwin 35, 50, 51
   Schrödinger equation 61, 66
Shoemaker, Sydney 119, 125, 129, 133
Sinai, Y. G. 168
Sklar, Lawrence 2, 99, 161, 178, 182, 197
Sneed, Joseph D. 139, 142
Sommerfeld radiation condition 60
space-time 120–131
Sparber, G. 122
Spohn, H. 39, 53, 199
statistical mechanics
   Boltzmannian statistical mechanics 92–95
   philosophical reflection on statistical mechanics 2
   statistical mechanical account 13, 16, 19, 20, 22, 26, 27, 28–29
   statistical mechanics and thermodynamics *see* reduction
Stegmüller, Wolfgang 71, 139

Stirzaker, D. R. 204
Stöltzner, M. 1
Streater, R. F. 181, 188
Strevens, Michael 77, 82, 85, 88, 90
string theory 34, 63
structuralism 139–140
Suárez, M. 131
subdynamics 200
Sudarshan, E. C. G. 180, 199

Taylor expansion 188
temperature
   absolute temperature 154
thermodynamics 2–3, 92
   phenomenological thermodynamics 37, 139, 142, 170
   second law of thermodynamics 3, 13, 15, 37, 46, 60, 92, 94, 123
   thermodynamic arrow of time *see* arrows of time
   thermodynamic asymmetry *see* Boltzmannian account of thermodynamic asymmetry
   thermodynamics and statistical mechanics *see* reduction
time
   direction of time 39, 42, 124, 127
   time-asymmetry 3, 4, 8, 13–14, 40, 181–182, 186, 200–201
   time-symmetry 15, 16, 34, 61, 183, 186, 190, 200–201
Touchette, H. 44
Tremaine, S. 45, 50
Tsallis statistics 47
   Tsallis entropy *see* entropy

Uffink, Jos 2, 8, 16, 48, 94, 95, 99
Universe
   entropy of the Universe *see* entropy
   expansion of Universe 3, 59–62
   macro-evolution of Universe 22

van Fraassen, Bas C. 132
van Kampen 54, 180, 182, 188, 192, 194, 195–196, 199, 201, 204
van Lith, Janneke 95, 99
Vienna Circle 83
Vlasov equation 52, 55
von Wright, G. H. 83

Waismann, Friedrich 83
Wald, R. M. 40, 42
wave function 61, 64, 66
Weber, T. 131
Wheeler–De Witt equation 4, 63–64
Wightman, A. S. 176
Winsberg, Eric 16, 101, 112
Wittgenstein, Ludwig 83

Yngvason, J. 46

Zeh, H. D. 59, 65, 66